SpringerBriefs in Regional Science

For further volumes:
http://www.springer.com/series/10096

Manfred M. Fischer · Jinfeng Wang

Spatial Data Analysis

Models, Methods and Techniques

 Springer

Prof. Dr. Manfred M. Fischer
SocioEconomics
Vienna University of Economics
and Business
Nordbergstraße 15/4
1090 Vienna
Austria
e-mail: manfred.fischer@wu.ac.at

Prof. Dr. Jinfeng Wang
State Key Laboratory of Resources and
Environmental Information Systems
Chinese Academy of Sciences
Datun Road 11A
100101 Beijing
People's Republic of China
e-mail: wangjf@lreis.ac.cn

ISSN 2192-0427
ISBN 978-3-642-21719-7
DOI 10.1007/978-3-642-21720-3
Springer Heidelberg Dordrecht London New York

e-ISSN 2192-0435
e-ISBN 978-3-642-21720-3

Cover design: eStudio Calamar, Berlin/Figueres

Printed on acid-free paper

Springer is part of Springer Science+Business Media (www.springer.com)

Preface

The centrality of space and location has always been taken as granted in geography and regional science. But recent attention to the spatial dimension of phenomena has also increased in the mainstream of the social sciences, and increasingly also in the natural sciences like ecology. A growing number of social scientists have taken up the use of new methodologies and technologies (such as geographic information systems, global positioning systems, remote sensing, spatial statistics and spatial econometrics) in the empirical work. In addition, increased attention is paid to location and spatial interaction in theoretical frameworks.

In broad terms, one might define spatial analysis as the quantitative analysis of spatial phenomena that are located in geographical space (Bailey and Gatrell 1995). It would be too ambitious to cover such a broad field in one textbook (see Fischer and Getis 2010 for an accounting of the diversity of the field). We thus decided to limit the scope to that important subset of spatial analysis which is known as *spatial data analysis*. In doing so, we are concerned with the situation where observational data are available on some process operating in geographic space, and consider models, methods and techniques to describe or explain the behaviour of this process and its possible relationship to other spatial phenomena. By defining spatial data analysis in this way we place the book in the area of statistical description and modelling of spatial data, and restrict ourselves to a particular set of methods. In doing so, we exclude some important quantitative methods such as, for example, various forms of network analysis and location-allocation analysis that would be included under the more general heading of spatial analysis.

Whether or not spatial data analysis is a separate academic field, the fact remains that in the last twenty years spatial data analysis has become an important by-product of the interest in and the need to understand spatial data. By spatial data we mean data which relate to observations with a spatial reference where spatial reference may be explicit, as in a postal address or a grid reference, or implicit, as a pixel in remote sensing.

The past decades have generated a number of excellent texts on the subject (see, for example, Cliff and Ord 1981; Upton and Fingleton 1985; Anselin 1988b;

Griffith 1988; Ripley 1988; Cressie 1993; Haining 1990, 2003; Bailey and Gatrell 1995; LeSage and Pace 2009). Most of these are addressed to the researcher. This text book is directed at introducing spatial data analysis to the graduate student, from a "data-driven" rather than a "theory-led" perspective. With this overall objective in mind, we have not attempted to discuss exhaustively the whole area of spatial data analysis, but have restricted the discussion to the analysis of two major types of spatial data: area data defined as data associated with a fixed set of areas or zones covering the study area, and spatial interaction (or origin–destination flow) data defined as measurements each of which associated with a link or pair of locations representing points or areas in geographical space.

We have restricted ourselves to a subset of models, methods and techniques which we believe to be relatively accessible and useful for analysing these types of spatial data. The topics discussed in this book include a mixture of both informal/ exploratory methods and techniques on the one side, and formal statistical modelling, parameter estimation and hypothesis testing on the other.

The book is divided into two parts. Each of these parts is as self-contained as possible. The first, Part I, considers the analysis of *area data*. The areas may form a regular lattice, as with remotely sensed images, or be a set of irregularly shaped areas, such as administrative districts. The second part, Part II, shifts attention to the analysis of spatial interaction data which are related to pairs of points or areas. Such data—called origin–destination flow or spatial interaction data—are relevant in studies of transport planning, population migration, journey-to-work, shopping behaviour, freight flows, and even the transformation of information and knowledge.

We do not consider spatiotemporal data, but assume that the data are purely spatial, either having been aggregated over time or referring to fixed points in time. Issues of measurement, storage and retrieval of spatial data are important, but outside the scope of this textbook. GISystems offer software tools that facilitate— through georeferencing—the integration of spatial and non-spatial, qualitative and quantitative data in a database that can be managed under one system environment (see Longley et al. 2001 for a discussion). In keeping the text in a manageable number of pages we assume our reader to have a moderate level of general background in statistics and mathematics.

We acknowledge the generous general support provided by the Institute for Economic Geography and GIScience, Vienna University of Economics and Business. We have benefitted greatly from the technical assistance Thomas Seyffertitz (Institute for Economic Geography and GIScience) provided. His expertise in handling several word processing systems, formatting, and indexing, together with his care and attention to detail, helped immeasurably. Last but not at least we thank the editor of the book series, Dr. Henk Folmer, for his valuable comments on an earlier version of the manuscript.

Vienna, Beijing, May 2011   Manfred M. Fischer
                                                                                        Jinfeng Wang

Contents

Chapter 1
Introduction

Abstract In this chapter we give an introduction to spatial data analysis, and distinguish it from other forms of data analysis. By spatial data we mean data that contain locational as well as attribute information. We focus on two broad types of spatial data: area data and origin–destination flow data. Area data relate to a situation where the variable of interest—at least as our book is concerned—does not vary continuously, but has values only within a fixed set of areas or zones covering the study area. These fixed sites may either constitute a regular lattice (such as pixels in remote sensing) or they may consist of irregular areal units (such as, for example, census tracts). Origin–destination flow (also called spatial inter-action) data are related instead to pairs of points, or pairs of areas in geographic space. Such data—that represent flows of people, commodities, capital, information or knowledge, from a set of origins to a set of destinations—are relevant in studies of transport planning, population migration, journey-to work, shopping behaviour, freight flows, and the transmission of information and knowledge across space. We consider the issue of spatial autocorrelation in the data, rendering conventional statistical analysis unsafe and requiring spatial analytical tools. This issue refers to situations where the observations are non-independent over space. And we conclude with a brief discussion of some practical problems which confront the spatial analyst.

Keywords Spatial data · Types of spatial data · Spatial data matrix · Area data · (Origin–destination) Flow data · Spatial autocorrelation · Tyranny of spatial data

1.1 Data and Spatial Data Analysis

Data consist of numbers, or symbols that are in some sense neutral and—in contrast to information—almost context-free. Raw geographical facts, such as the temperature at a specific time and location, are examples of data. Following

M. M. Fischer and J. Wang, *Spatial Data Analysis*,
SpringerBriefs in Regional Science, DOI: 10.1007/978-3-642-21720-3_1,
© Manfred M. Fischer 2011

Longley et al. (2001, p. 64) we can view spatial data as built up from atomic elements or facts about the geographic world. At its most primitive, an atom of spatial data (strictly, a datum) links a geographic location (place), often a time, and some descriptive property or attribute of the entity with each other. For example, consider the statement "The temperature at 2 pm on December 24, 2010 at latitude 48°15′ North, longitude 16°21′ 28 s East, was 6.7°C". It ties location and time to the property or attribute of atmospheric temperature. Hence, we can say that spatial (geographic) data link place (location), time and an attribute (here: temperature).

Attributes come in many forms. Some are physical or environmental in nature, while others are social or economic. Some simply identify a location such as postal addresses or parcel identifiers used for recording land ownership. Other attributes measure something at a location (examples include atmospheric temperature and income), while others classify into categories such as, for example, land use classes that differentiate between agriculture, residential land and industry.

While time is optional in spatial data analysis, geographic location is essential and distinguishes *spatial data analysis* from other forms of data analysis that are said to be non-spatial or aspatial. If we would deal with attributes alone, ignoring the spatial relationships between sample locations, we could not claim of doing spatial data analysis, even though the observational units may themselves be spatially defined. Even if attribute data would be of fundamental importance, divorced from their spatial context, they lose value and meaning (Bailey and Gatrell 1995, p. 20). In order to undertake spatial data analysis, we require—as a minimum—information for both locations and attributes, regardless, of how the attributes are measured.

Spatial data analysis requires an underlying spatial framework on which to locate the spatial phenomena under study. Longley et al. (2001) and others have drawn a distinction between two fundamental ways of representing geography: a *discrete* and a *continuous* view of spatial phenomena. In other words, a distinction is made between a conception of space as something filled with "discrete objects", and a view of space as covered with essentially "continuous surfaces". The former has been labelled an *object* or *entity view of space*, the latter a *field view*.

In the object view the sorts of spatial phenomena being analysed are identified by their dimensionality. Objects that occupy area are called two-dimensional, and are generally referred to as areas. Other objects are more like one-dimensional lines, including rivers, railways, or roads, and are represented as one-dimensional objects and generally referred to as lines. Other objects are more like zero-dimensional points, such as individual plants, people, buildings, the epicentres of earthquakes, and so on, and are referred to as points (Longley et al. 2001, pp. 67–68; Haining 2003, pp. 44–46). Note that surface or volume objects—not considered in this book—have length, breadth, and depth, and thus are three-dimensional. They are used to represent natural objects such as river basins or artificial phenomena such as the population potential of shopping centres.

Of course, how appropriate this is depends upon the spatial scale (level of detail at which we seek to represent "reality") of study. If we are looking at the distribution of urban settlements at a national scale, it is reasonable to treat them as a distribution of points. At the scale of a smaller region, for example, it becomes less sensitive. Phenomena such as roads can be treated as lines as mentioned above. But there is again scale dependence. On large scale maps of urban areas roads have a width, and this may be important when interest is on car navigation issues, for example. Lines also mark the boundaries of areas. By areas we generally understand those entities which are administratively or legally defined, such as countries, districts, census zones, and so on, but also "natural areas" such as soil or vegetation zones on a map.

In a field view the emphasis is on the continuity of spatial phenomena, and the geographic world is described by a finite number of variables, each measurable at any point of the earth's surface, and changing in value across the surface (Haining 2003, pp. 44–45). If we think of phenomena in the natural environment such as temperature, soil characteristics, and so on, then such variables can be observed anywhere on the earth's surface (Longley et al. 2001, pp. 68–71). Of course, in practice such variables are discretised. Temperature, for example, is sampled at a set of sites and represented as a collection of lines (so-called isotherms). Soil characteristics might be also sampled at a set of discrete locations and represented as a continuously varying field. In all such cases, an attempt is made to represent underlying continuity from discrete sampling (Bailey and Gatrell 1995, p. 19).

1.2 Types of Spatial Data

In describing the nature of spatial data it is important to distinguish between the discreteness or continuity of the space on which the variables are measured, and the discreteness or continuity of the variable values (measurements) themselves. If the space is continuous (a field view), variable values must be continuous valued since continuity of the field could not be preserved under discrete valued variables. If the space is discrete (an object view) or if a continuous space has been made discrete, variable values may be continuously valued or discrete valued (nominal or ordinal valued) (see Haining 2003, p. 57).

The classification of spatial data by type of conception of space and level of measurement is a necessary first step in specifying the appropriate statistical technique to use to answer a question. But the classification is not sufficient because the same spatial object may be representing quite different geographical spaces. For example, points (so-called centroids) are also used to represent areas. Table 1.1 provides a typology that distinguishes four types of spatial data:

(i) *point pattern data*, that is, a data set consisting of a series of point locations in some study region, at which events of interest (in a general sense) have occurred, such as cases of a disease or incidence of a type of crime,

(ii) *field data* (also termed *geostatistical data*) that relate to variables which are conceptually continuous (the field view) and whose observations have been sampled at a predefined and fixed set of point locations,

Table 1.1 Types of spatial data: conceptual schemes and examples

Type of spatial data	Conceptual scheme		Example		Space
	Variable scheme	Spatial index	Variable		Space
Point pattern data	Variable (discrete or continuous) is a random variable	Point objects to which the variable is attached are fixed	Trees: diseased or not Hill forts: classified by type		Two-dimensional discrete space Two-dimensional discrete space
Spatially continuous (geostatistical) data	Variable is a continuous valued function of location	Variable is defined everywhere in the (two-dimensional) space	Temperature Atmospheric pollution		Two-dimensional continuous space Two-dimensional continuous space
Area (object) data	Variable (discrete or continuous) is a random variable	Area objects to which the variable is attached are fixed	Gross regional product Crime rates		Two-dimensional discrete space Two-dimensional discrete space
Spatial interaction (flow) data	Variable representing mean interaction frequencies is a random variable	Pairs of locations (points or areas) to which the flow variable is attached	International trade Population migration		Two-dimensional discrete space Two-dimensional discrete space

(iii) *area data* where data values are observations associated with a fixed number of areal units (area objects) that may form a regular lattice, as with remotely sensed images, or be a set of irregular areas or zones, such as counties, districts, census zones, and even countries,

(iv) *spatial interaction data* (also termed *origin–destination flow* or *link data*), consisting of measurements each of which is associated with a pair of point locations, or pair of areas.

In this book, we do neither consider point pattern data nor field (geostatistical) data. The focus is rather on the analysis of object data where the observations relate to areal units (see Part I) and on the analysis of origin–destination flow (spatial interaction) data (see Part II). The analysis of spatial interaction data has a long and distinguished history in the study of human activities, such as transportation movements, migration, and the transmission of information and knowledge. And area data provide an important perspective for spatial data analysis applications, in particular in the social sciences.

1.3 The Spatial Data Matrix

All the analytical techniques in this book use a data matrix that captures the spatial data needed for the conduct of analysis. Spatial data are classified by the type of spatial object (point object, area object) to which variables refer and the level of measurement of these variables.

Let $Z_1, Z_2, \ldots, Z_K$ refer to K random variables and S to the location of the point or area. Then the spatial data matrix (see Haining 2003, pp. 54–57) can be generally represented as

$$
\begin{array}{cccc}
\text{Data on the } K \text{ variables} & & \text{Location} \\
Z_1 \quad Z_2 \quad \ldots \quad Z_K & S
\end{array}
$$

$$
\begin{bmatrix}
z_1(1) & z_2(1) & \ldots & z_K(1) & s(1) \\
z_1(2) & z_2(2) & \ldots & z_K(2) & s(2) \\
\vdots & \vdots & & \vdots & \vdots \\
z_1(n) & z_2(n) & \ldots & z_K(n) & s(n)
\end{bmatrix}
\begin{array}{l}
\text{Case 1} \\
\text{Case 2} \\
\vdots \\
\text{Case } n
\end{array}
$$

which may be shortened to

$$
\left\{ z_1(i), z_2(i), \ldots, z_K(i) \mid s(i) \right\}_{i=1,\ldots,n} \tag{1.1}
$$

where the lower case symbol z_k denotes an realisation (actual data value) of variable Z_k $(k = 1, \ldots, K)$ while the symbol i inside the brackets references the particular case. Attached to each case $i = 1, \ldots, n$ is a location $s(i)$ that represents the location of the spatial object (point or area). Since we are only interested in *(a)* Assigning locations to point objects

Case i	$s(i)$		Variables			
	s_1	s_2	Z_1	Z_2	$\dots$	Z_K
1	$s_1(1)$	$s_2(1)$	$z_1(1)$	$z_2(1)$	$\dots$	$z_K(1)$
2	$s_1(2)$	$s_2(2)$	$z_1(2)$	$z_2(2)$	$\dots$	$z_K(2)$
$\vdots$	$\vdots$	$\vdots$	$\vdots$	$\vdots$		$\vdots$
n	$s_1(n)$	$s_2(n)$	$z_1(n)$	$z_2(n)$	$\dots$	$z_K(n)$

(b) Assigning locations to irregularly shaped area objects

Case i	$s(i)$	Variables			
		Z_1	Z_2	$\dots$	Z_K
1	1	$z_1(1)$	$z_2(1)$	$\dots$	$z_K(1)$
2	2	$z_1(2)$	$z_2(2)$	$\dots$	$z_K(2)$
$\vdots$	$\vdots$	$\vdots$	$\vdots$		$\vdots$
n	n	$z_1(n)$	$z_2(n)$	$\dots$	$z_K(n)$

$1, 2, \dots, n$
look up
table

x denotes centroid

(c) Assigning locations to regularly shaped area objects

Case i	$s(i)$		Variables			
	p	q	Z_1	Z_2	$\dots$	Z_K
1	$s_1(1)$	$s_2(1)$	$z_1(1)$	$z_2(1)$	$\dots$	$z_K(1)$
2	$s_1(2)$	$s_2(2)$	$z_1(2)$	$z_2(2)$	$\dots$	$z_K(2)$
$\vdots$	$\vdots$	$\vdots$	$\vdots$	$\vdots$		$\vdots$
n	$s_1(n)$	$s_2(n)$	$z_1(n)$	$z_2(n)$	$\dots$	$z_K(n)$

Fig. 1.1 Assigning locations to spatial objects (points, areas) (adapted from Haining 2003, p. 55)

two-dimensional space, referencing will involve two geographic coordinates s_1 and s_2. Thus, $s(i) = (s_1(i), s_2(i))'$ where $(s_1(i), s_2(i))'$ is the transposed vector of $(s_1(i), s_2(i))$. It is important to note that in this book we generally consider methods that treat locations as fixed and do not consider problems where there is a randomness associated with the location of the cases.

In the case of data referring to point objects in two-dimensional space the location of the ith point may be given by a pair of (orthogonal) Cartesian coordinates as illustrated in Fig. 1.1a. The axes of the coordinate system will usually have been constructed for the particular data set, but a national or global referencing system may be used. In the case of data referring to irregularly shaped area objects one option is to select a representative point such as the centroid and then use the same procedure as for a point object to identify $s(i) = (s_1(i), s_2(i))'$ for $i = 1, \dots, n$. Alternatively, each area i is labelled and a look-up table provided so that rows of the data matrix can be matched to areas on the map (see Fig. 1.1b). If the areas are regularly shaped as in the case of a remotely sensed image they may be labelled as in Fig. 1.1c.

There are situations where the georeferencing information provided by $\{s(i)\}$ in expression (1.1) has to be supplemented with neighbourhood information that defines not only which pairs of areas are adjacent to each other but may also quantify the closeness of that adjacency. This information is needed for the specification of many spatial statistical models such as spatial regression models.

It is worth noting that on various occasions throughout the book, the variables $Z_1, \ldots, Z_K$ will be divided into groups and labelled differently. In the case of data modelling, the variable whose variation is to be modelled will be denoted Y and the variables used to explain the variations in the dependent variable are called explanatory or independent variables, labelled differently such as $X_1, \ldots, X_Q$.

Spatial interaction data record flows between locations (points, areas) or between nodes (intersection points) of a network. The situation, we are considering in this book is one of a series of observations $y_{ij}(i, j = 1, \ldots, n)$, on random variables Y_{ij}, each of which corresponds to movements of people, goods, capital, information, knowledge, and so on between spatial locations i and j, where these locations may be point locations or alternatively areas or zones. These data are captured in the form of an origin–destination or spatial interaction matrix

Destination location

$$\text{Origin location} \begin{bmatrix} y_{11} & y_{12} & \cdots & y_{1n} \\ y_{21} & y_{22} & \cdots & y_{2n} \\ \vdots & \vdots & & \vdots \\ y_{n'1} & y_{n'2} & \cdots & y_{n'n} \end{bmatrix} \tag{1.2}$$

where the number of rows and columns correspond to the number of origin and destination locations, respectively, and the entry on row i and column j, y_{ij}, records the observed total flow from origin location i to destination location j. In the special case where each location is both origin and destination $n' = n$. Georeferencing of the origin and destination locations follows the same procedures as described in the above case of object data.

1.4 Spatial Autocorrelation

The basic tenet underlying the analysis of spatial data is the proposition that values of a variable in near-by locations are more similar or related than values in locations that are far apart. This inverse relation between value association and distance is summarised by Tobler's first law stating that *"everything is related to everything else, but near things are more related than distant things"* (Tobler 1970, p. 234).

If near-by observations (i.e. similar in location) are also similar in variable values then the pattern as a whole is said to exhibit *positive* spatial autocorrelation (self-correlation). Conversely, *negative* spatial autocorrelation is said to exist when observations that are near-by in space tend to be more dissimilar in variable values than observations that are further apart (in contradiction to Tobler's law). Zero autocorrelation occurs when variable values are independent of location. It is important to note that spatial autocorrelation renders conventional statistical analysis invalid and makes spatial data analysis different from other forms of data analysis.

A crucial aspect of defining spatial autocorrelation is the determination of nearby locations, that is, those locations surrounding a given data point that could be considered to influence the observation at that data point. Unfortunately, the determination of that neighbourhood is not without some degree of arbitrariness.

Formally, the membership of observations in the neighbourhood set for each location may be expressed by means of an n-by-n spatial contiguity or weights matrix W

$$
W = \begin{bmatrix}
W_{11} & W_{12} & \cdots & W_{1n} \\
W_{21} & W_{22} & \cdots & W_{2n} \\
\vdots & \vdots & & \vdots \\
W_{n1} & W_{n2} & \cdots & W_{nn}
\end{bmatrix}
\tag{1.3}
$$

where n represents the number of locations (observations). The entry on row i ($i = 1, \ldots, n$) and column j ($j = 1, \ldots, n$), denoted as W_{ij}, corresponds to the pair (i, j) of locations. The diagonal elements of the matrix are set to zero, by convention, while the non-diagonal elements W_{ij} ($i \neq j$) take on non-zero values (one, for a binary matrix) when locations i and j are considered to be neighbours, otherwise zero.

For areal objects, such as the simple nine-zone system shown in Fig. 1.2a, (first order) spatial contiguity (or adjacency) is often used to specify neighbouring locations in the sense of sharing a common border. On this basis, Fig. 1.2a may be re-expressed as the graph shown in Fig. 1.2b. Coding $W_{ij} = 1$ if zones i and j are contiguous, and $W_{ij} = 0$ otherwise, we may derive a weights matrix W shown in Table 1.2. This matrix provides an example of the simplest way of specifying W.

In the classical case of a regular square grid layout the options of contiguity are referred to as the *rook* contiguity case (only common boundaries), the *bishop* contiguity case (only common vertices), and the *queen* contiguity case (both boundaries and vertices). Depending on the chosen criterion, an area will have four (rook, bishop) or eight (queen) neighbours on average. This implies quite different neighbour structures. Even in the case of irregularly shaped areal units, a decision has to be made whether areas that only share a common vertex should be considered to be neighbours (queen criterion) or not (rook criterion).

Contiguity may and is often defined as a function of the distance between locations (areas, points). In this sense, two objects are considered to be contiguous if the distance between them falls within a chosen range. In essence, the spatial weights matrix summarises the topology of the data set in graph-theoretic terms (nodes and links).

Higher order contiguity is defined in a recursive manner, in the sense that an object (point, area) is considered to be contiguous of a higher order to a given object if it is

Fig. 1.2 A zoning system:
a a simple mosaic of discrete
zones, **b** re-expressed as a
graph

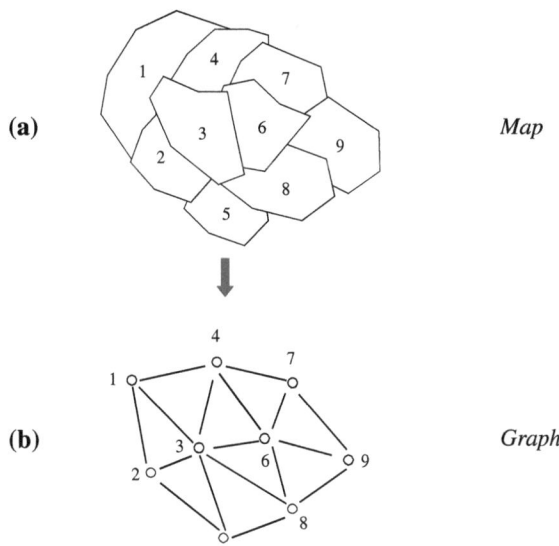

(a)   *Map*

(b)   *Graph*

Table 1.2 A spatial weights matrix W derived from the zoning system in Fig. 1.2: the case of a binary first order contiguity matrix

	1	2	3	4	5	6	7	8	9
1	0	1	1	1	0	0	0	0	0
2	1	0	1	0	1	0	0	0	0
3	1	1	0	1	1	1	0	1	0
4	1	0	1	0	0	1	1	0	0
5	0	1	1	0	0	0	0	1	0
6	0	0	1	1	0	0	1	1	1
7	0	0	0	1	0	1	0	0	1
8	0	0	1	0	1	1	0	0	1
9	0	0	0	0	0	1	1	1	0

first order contiguous to an object that is contiguous to an object that is contiguous of the next lower order. For example, objects that are viewed to be second order contiguous to an object are first order contiguous to the first order contiguous ones. In Fig. 1.2a, for example, areas 1 and 2 are first order contiguous to area 3, and area 3 is first order contiguous to area 6. Hence, areas 1 and 2 are second order contiguous to area 6. Thus, higher order contiguity yields bands of observations around a given location being included in the neighbourhood set, at increasing instances.

Clearly, a large number of spatial weights matrices may be derived for a given spatial layout such as that one shown in Fig. 1.2a. In particular, the spatial weights matrix does not have to be binary, but can take on any value that reflects the interaction between spatial units i and j, for example, based on inverse distances or inverse distances raised to some power.

The type of matrix shown in Table 1.2 allows us to develop measures of spatial autocorrelation. Many tests and indicators of spatial autocorrelation are available.

Chief among these is Moran's spatial autocorrelation statistic (see Cliff and Ord 1973, 1981). At the local scale Getis and Ord's statistics (see Getis and Ord 1992; Ord and Getis 1995) and Anselin's LISA statistics (see Anselin 1995) enable analysts to evaluate spatial autocorrelation at particular sites. We will say more about this in the next chapter.

1.5 The Tyranny of Spatial Data

Spatial data analysis crucially depends on data quality. Good data are reliable, contain few or no mistakes, and can be used with confidence. Unfortunately, nearly all spatial data are flawed to some degree. Errors may arise in measuring both the location (points, lines, areas) and attribute properties of spatial objects. In the case of measurements of location (position), for example, it is possible for every coordinate to be subject to error. In the two-dimensional case, a measured location would be subject to error in both coordinates.

Attribute errors can arise as a result of collecting, storing, manipulating, editing or retrieving attribute values. They can also arise from inherent uncertainties associated with the measurement process and definitional problems, including the point or area location a measurement refers to (Haining 2003, pp. 59–63; see also Wang et al. 2010). The solution to the data quality problem is to take the necessary steps to avoid having faulty data determining research results.

The particular form (i.e. size, shape and configuration) of the spatial aggregates can affect the results of the analysis to a varying—usually unknown—degree as evidenced in various types of analysis (see, for example, Openshaw and Taylor 1979; Baumann et al. 1983). This problem generally has become recognised as the *modifiable areal unit problem* (MAUP), the term stemming from the fact that areal units are not 'natural' but usually arbitrary constructs.

Confidentiality restrictions usually dictate that data (for example, census data) may not be released for the primary units of observation (individuals, households or firms), but only for a set of rather arbitrary areal aggregations (enumeration districts or census tracts). The problem arises whenever area data are analysed or modelled and involves two effects: One derives from selecting different areal boundaries while holding the overall size and the number of areal units constant (the zoning effect). The other derives from reducing the number but increasing the size of the areal units (the scale effect). There is no analytical solution to the MAUP (Openshaw 1981), but the modifiable areal unit problem can be investigated through simulation of large numbers of alternative systems of areal units (Longley et al. 2001, p. 139). Such systems can obviously take many different forms, both in relation to the level of spatial resolution and also in relation to the shape of the areas.

An issue that has been receiving increasing attention in recent years relates to the data suitability problem. It is not unusual to find published work where the researcher uses data available at one spatial scale to come to conclusions about a relationship or process at a finer scale. This *ecological fallacy*, as it is known, leads us into a false sense of the power of our techniques and usefulness of our conclusions (Getis 1995). The ecological fallacy and the modifiable areal unit problem have long been recognised as problems in applied spatial data analysis, and, through the concept of spatial autocorrelation, they are understood as related problems.

Part I
The Analysis of Area Data

In Part I we consider the analysis of area data. Area data are observations associated with a fixed number of areal units (areas). The areas may form a regular lattice, as with remotely sensed images, or be a set of irregular areas or zones, such as countries, districts and census zones.

We draw a distinction between methods that are essentially exploratory in nature, concerned with mapping and geovisualisation, summarising and analysing patterns, and those which rely on the specification of a statistical model and the estimation of the model parameters. The distinction is useful, but not clear cut. In particular, there is usually a close interplay of the two, with data being visualised and interesting aspects being explored, which possibly lead to some modelling.

Chapter 2 will be devoted to methods and techniques of exploratory data analysis that concentrates on the spatial aspects of the data, that is, exploratory spatial data analysis [ESDA] (see Haining 2003; Bivand 2010). The focus is on univariate techniques that elicit information about spatial patterns of a variable, and identify atypical observations (outliers).

Exploratory spatial data analysis is often only a preliminary step towards more formal modelling approaches that seek to establish relationships between the observations of a variable and the observations of other variables, recorded for each area. Chapter 3 provides a concise overview of some of the central methodological issues related to spatial regression analysis in a simple cross-sectional setting.

Keywords Area data · Spatial weights matrix · Moran's I statistic · Geary's c statistic · G statistics · LISA statistics · Spatial regression models · Spatial Durbin model · Tests for spatial dependence · Maximum likelihood estimation · Model parameter interpretation

Chapter 2
Exploring Area Data

Abstract Here in this chapter, we first consider the visualisation of area data before examining a number of exploratory techniques. The focus is on spatial dependence (spatial association). In other words, the techniques we consider aim to describe spatial distributions, discover patterns of spatial clustering, and identify atypical observations (outliers). Techniques and measures of spatial autocorrelation discussed in this chapter are available in a variety of software packages. Perhaps the most comprehensive is GeoDa, a free software program (downloadable from http://www.geoda.uiuc.edu). This software makes a number of exploratory spatial data analysis (ESDA) procedures available that enable the user to elicit information about spatial patterns in the data given. Graphical and mapping procedures allow for detailed analysis of global and local spatial autocorrelation results. Another valuable open software is the spdep package of the R project (downloadable from http://cran.r-project.org). This package contains a collection of useful functions to create spatial weights matrix objects from polygon contiguities, and various tests for global and spatial autocorrelation (see Bivand et al. 2008).

Keywords Area data · Spatial weights matrix · Contiguity-based specifications of the spatial weights matrix · Distance-based specifications of the spatial weights matrix · k-nearest neighbours · Global measures of spatial autocorrelation · Moran's I statistic · Geary's c statistic · Local measures of spatial autocorrelation · G statistics · LISA statistics

2.1 Mapping and Geovisualisation

In exploratory spatial data analysis the map has an important role to play. The map is the most established and conventional means of displaying areal data. There is a variety of ways ascribing continuous variable data to given areal units that are predefined. In practice, however, none is unproblematic. Perhaps the most commonly

M. M. Fischer and J. Wang, *Spatial Data Analysis*,
SpringerBriefs in Regional Science, DOI: 10.1007/978-3-642-21720-3_2,

used form of display is the standard *choropleth* map (Longley et al. 2001, pp. 251–252; Bailey and Gatrell 1995, pp. 255–260; Demšar 2009, pp. 48–55). This is a map where each of the areas is coloured or shaded according to a discrete scale based on the value of the variable (attribute) of interest within that area. The number of classes (categories) and the corresponding class (category) intervals can be based on several different criteria.

There are no hard rules about numbers of classes. Clearly, this is a function of how many observations we have. For example, if we have only a sample of 20 or 30 areas it hardly makes sense to use seven or eight classes. But perhaps with some hundreds of measurements a set of seven or eight classes is likely to prove informative. As a general rule of thumb some statisticians recommend a number of classes of $(1 + 3.3 \ln n)$, where n is the number of areas and 'ln' stands for the logarithm naturalis (Bailey and Gatrell 1995, p. 153).

As for class interval selection, four basic classification schemes may be used to divide interval and ratio areal data into categories (Longley et al. 2001, p. 259):

(i) *Natural breaks* by which classes are defined according to some natural groupings of the data values. The breaks may be imposed on the basis of break points which are known to be relevant in a particular application context, such as fractions and multiples of mean income levels, or rainfall thresholds of vegetation ('arid', 'semi-arid', 'temperate' etc.). This is a deductive assignment of breaks, while inductive classifications of data values may be carried out by using GISystem software tools to look for relatively large jumps in data values, as shown in Fig. 2.1a.

(ii) *Quantile breaks*, where each of a predetermined number of classes (categories) contains an equal number of observations (see Fig. 2.1b). Quartile (four category) and quintile (five category) classifications are commonly used in practice. The numeric size of each class is rigidly imposed. Note that the placing of the class boundaries may assign almost identical observations to adjacent classes, and observations with quite widely different values to the same class. The resulting visual distortion can be minimised by increasing the number of classes.

(iii) *Equal interval breaks* are self-explanatory (see Fig. 2.1c). They are valuable where observations are reasonably uniformly distributed over their range. But if the data are markedly skewed they will give large numbers of observations in just a few classes. This is not necessarily a problem, since unusually high (low) values are easily picked out on the map. An extension of this scheme is to use "trimmed equal" intervals where the top and bottom of the frequency distribution (for example, the top and bottom ten percent) are each treated as separate classes and the remainder of the observations are divided into equal classes.

(iv) *Standard deviation classifications* are based on intervals distributed around the mean in units of standard deviation (see Fig. 2.1d). They show the distance of an observation from the mean. One calculates the mean value and then generates class breaks in standard deviation measures above and below it.

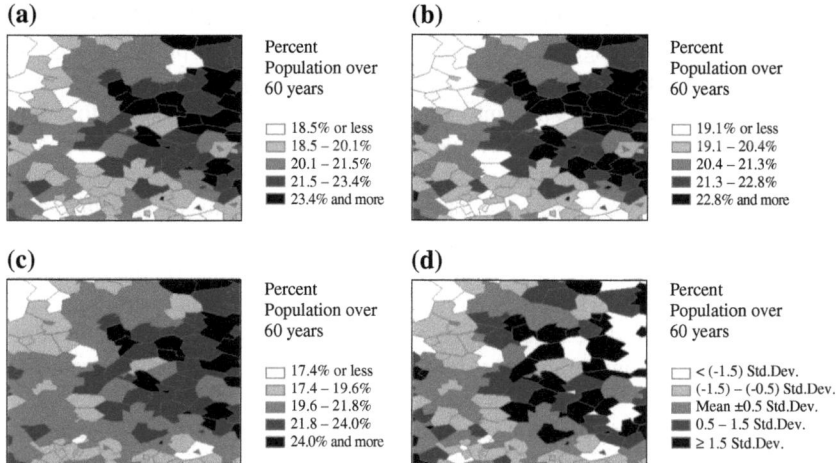

Fig. 2.1 Class definition using **a** natural breaks, **b** quantile breaks, **c** equal interval breaks, and **d** standard deviation breaks

We can obtain a large variety of maps simply by varying the class intervals. It is important to try out some of the above possibilities to get some initial sense of spatial variation in the data.

There are also other problems associated with the use of choropleth maps. *First*, choropleth maps bring the (dubious) visual implication of within-area uniformity of variable values. Moreover, conventional choropleth mapping allows any physically large area to dominate the display, in a way which may be quite inappropriate for the type of data being mapped. For example, in mapping socioeconomic data, large and sparsely populated rural areas may dominate the choropleth map because of the visual 'intrusiveness' of the large areas. But the real interest may be in physically smaller areas, such as the more densely populated urban areas.

A variant of the conventional choropleth map is the dot density map that uses dots as a more aesthetically pleasing means of representing the relative density of zonally averaged data, but not as a means of depicting the precise location of point events. Proportional circles provide one way around this problem, since the circle can be centred on any convenient point within an areal unit. But there is a tension between using circles that are of sufficient size to convey the variability in the data and the problems of overlapping circles (Longley et al. 2001, p. 259).

Second, the variable of interest has arisen from the aggregation of individual data to the areas. It has to be taken into account that these areas may have been designed rather arbitrarily on the basis of administrative convenience or ease of enumeration. Hence, any pattern that is observed across the areas may be as much a function of the area boundaries chosen, as it is of the underlying spatial distribution of variable values. This has become known as the *modifiable areal unit problem*. It can be a particularly significant problem in the analysis of socioeconomic and demographic data, where the enumeration areas have rarely been arrived at any basis that relates to the data under study (see also Sect. 1.5).

A solution to the problem of modifiable areal units is difficult. The ideal solution is to avoid using area aggregated data altogether if at all possible. In some applications of spatial data analysis, such as epidemiology and crime, one could perform valuable analyses on point data, without aggregating the data to a set of inherently arbitrary areal units. But of course in many cases such an approach will be not viable, and one has to live with areal units for which data are available.

Third, it is important to realise that the statistical results of any analysis of patterns and relationships will inevitably depend on the particular areal configuration which is being used. In general, data should be analysed on the basis of the smallest areal units for which they are available and aggregation to arbitrary larger areas should be avoided, unless there are good reasons to doing so. It is also important to check any inferences drawn from the data by using different areal configurations of the same data, if possible.

One approach to the problem of the dominance of physically large areas is to geometrically transform each of the areal units in such a way as to make its area proportional to the corresponding variable value, whilst at the same time maintaining the spatial contiguity of the areal units. The resulting map is often termed *cartogram* (Bailey and Gatrell 1995, p. 258). Cartograms lack planimetric correctness, and distort area or distance in the interest of some specific objective. The usual objective is to reveal patterns that might not be readily apparent from a conventional map. Thus, the integrity of the spatial object (area), in terms of areal extent, location, contiguity, geometry, and topology is made subservient to an emphasis upon variable values or particular aspects of spatial relations.

An example of a cartogram is given in Fig. 2.2 that shows a country's size as the proportion of global gross domestic product (*gdp*) found there in 2005, measured in terms of constant US dollars. The map reveals that global *gdp* is concentrated in a few world regions, in North America, Western Europe and North-East Asia. This global concentration matters greatly for the development prospects of today's lagging world regions, especially Africa which shows up as a slender peninsula in this cartogram.

Mapping and geovisualisation is an important step to provoke questions, but exploratory data analysis requires highly interactive, dynamic data displays. Recent developments in spatial data analysis software provide an interactive environment that combines maps with statistical graphs, using the technology of dynamically linked windows. Perhaps, the most comprehensive software with such capabilities is GeoDa. GeoDa includes functionality from conventional mapping to exploratory data analytic tools, and the visualisation of global and local autocorrelation. The software adheres to ESRI's (Environmental Systems Research Institute's) shape file as the standard for storing spatial information, and uses ESRI's Map-Objects LT2 technology for spatial data access, mapping and querying.

All graphic windows are based on Microsoft Foundation Classes and hence limited to Microsoft Windows platforms. In contrast, the computational engine (including statistical operations) is pure C++ code and largely cross platform. The bulk of the graphical interface implements five basic classes of windows:

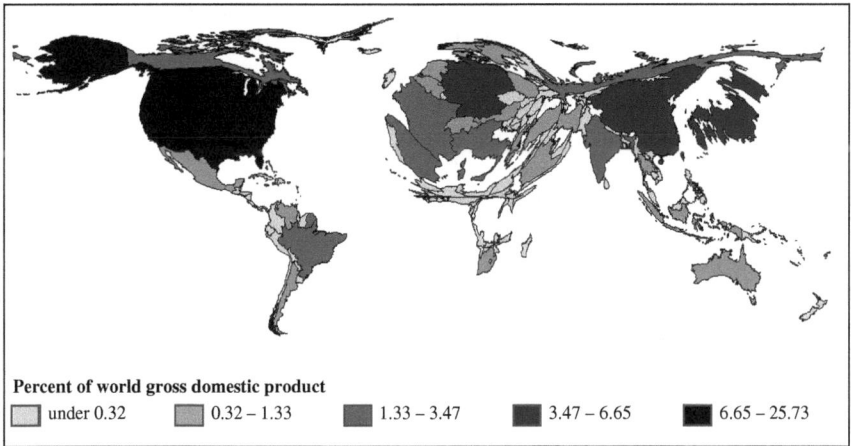

Fig. 2.2 A cartogram illustrating the concentration of global gross domestic product in a few world regions. A country's size shows the proportion of global gross domestic product found there. *Data source:* GISCO-Eurostat (European Commission); *Copyright:* EuroGeographics for the European administrative boundaries; *Copyright:* UN-FAO for the world administrative boundaries (except EuroGeographics members)

map, histogram, box plot, scatter plot (including the Moran scatter plot, see Anselin 1996), and grid (for the table selection and calculations). The choropleth map, including cluster maps for the local indicators of spatial autocorrelation (see Sect. 2.4), are derived from MapObjects classes. For an outline of the design and review of the overall functionality of GeoDa see Anselin et al. (2010).

2.2 The Spatial Weights Matrix

The focus of exploratory spatial data analysis is on measuring and displaying global and local patterns of spatial association, indicating local non-stationarity, discovering islands of spatial heterogeneity, and so on. A crucial aspect of defining spatial association (autocorrelation) is the determination of the relevant "neighbourhood" of a given area, that is, those areal units surrounding a given data point (area) that would be considered to influence the observation at that data point. In other words, neighbouring areas are spatial units that interact in a meaningful way. This interaction could relate, for example, to spatial spillovers and externalities.

The neighbourhood structure of a data set is most conveniently formalised in form of a spatial weights matrix W, of dimension equal to the a priori given number n of areal units considered. Each area is identified with a point (centroid) where Cartesian coordinates are known. In this matrix each row and matching column corresponds to an observation pair. The elements W_{ij} of this matrix take on a non-zero value (one for a binary matrix) when areas (observations) i and j are considered to be neighbours, and a zero value otherwise. By convention, an observation is not a neighbour to itself, so that the main diagonal elements W_{ii} ($i = 1, \ldots, n$) are zero.

The spatial weights matrix is often row-standardised, that is, each row sum in the matrix is made equal to one, the individual values W_{ij} are proportionally represented. Row-standardisation of W is desirable so that each neighbour of an area is given equal weight and the sum of all W_{ij} (over j) is equal to one. If the observations are represented as an n-by-1 vector X, the product, WX, of such a row-standardised weights matrix W with X has an intuitive interpretation. Since for each element i WX equals $\Sigma_j W_{ij}X_j$, WX is in fact a vector of weighted averages of neighbouring values. This operation and the associated variable are typically referred to as (first order) spatial lag of X, similar to the terminology used in time series analysis. Note that the space in which the observations are located need not to be geographic, any type of space is acceptable as long as the analyst can specify the spatial interactions between the areas given.

One way to represent the spatial relationships with areal data is through the concept of contiguity. First order contiguous neighbours are defined as areas that have a common boundary. Formally,

$$W_{ij} = \begin{cases} 1 & \text{if area } j \text{ shares a common boundary with area } i \\ 0 & \text{otherwise.} \end{cases} \tag{2.1}$$

Alternatively, two areas i and j may be defined as neighbours when the distance d_{ij} between their centroids is less than a given critical value, say d, yielding distance-based spatial weights

$$W_{ij} = \begin{cases} 1 & \text{if } d_{ij} < d \, (d > 0) \\ 0 & \text{otherwise} \end{cases} \tag{2.2}$$

where distances are calculated from information on latitude and longitude of the centroid locations. Examples include straight-line distances, great circle distances, travel distances or times, and other spatial separation measures.

Straight-line distances determine the shortest distance between any two point locations in a flat plane, treating longitude and latitude of a location as if they were equivalent to plane coordinates. In contrast, great circle distances determine distances between any two points on a spherical surface such as the earth as the length of the arc of the great circle between them (see Longley et al. 2001, pp. 86–92, for more details). In many applications the simple measures—straight-line distances and great circle distances—are not sufficiently accurate estimates of actual travel distances, and one is forced to resort to summing the actual lengths of travel routes, using a GISystem. This normally means summing the length of links in a network representation of a transportation system.

The distance-based specification (2.2) of the weights matrix depends on a given critical distance value, d. When there is a high degree of heterogeneity in the size of the areal units, however, it can be difficult to find a satisfactory critical distance. In such circumstances, a small distance will tend to lead to a lot of islands (i.e. unconnected observations), while a distance chosen to guarantee that each areal unit (observation) has at least one neighbour may yield an unacceptably large number of neighbours for the smaller areal units (Anselin 2003a).

In empirical applications, this problem is encountered when building distance-based spatial weights, for example, for NUTS-2 regions in Europe, where such areal units in sparsely populated parts of Europe are much larger in physical size than in more populated parts such as in Central Europe. A common solution to this problem is to constrain the neighbour structure to the *k-nearest* neighbours, and thereby precluding islands and forcing each areal unit to have the same number k of neighbours. Formally,

$$W_{ij} = \begin{cases} 1 & \text{if centroid of } j \text{ is one of the } k \text{ nearest centroids to that of } i \\ 0 & \text{otherwise.} \end{cases} \tag{2.3}$$

If the number of nearest neighbours, for example, is set to six, then the non-normalised weights matrix will have six ones in each row, indicating the six closest observations to $i = 1, \ldots, n$. The number of neighbours, k, is the parameter of this weighting scheme. The choice of k remains an empirical matter (see LeSage and Fischer 2008).

The above specifications of the spatial weights matrix share the property that their elements are fixed. It is straightforward to extend this notion by changing the weighting on the neighbours so that more distant neighbours get less weight by introducing a parameter θ that allows to indicate the rate of decline of the weights. A commonly used continuous weighting scheme is based on the inverse distance function so that the weights are inversely related to the distance separating area i and area j

$$W_{ij} = \begin{cases} d_{ij}^{-\theta} & \text{if intercentroid distance } d_{ij} < d \, (d > 0, \theta > 0) \\ 0 & \text{otherwise} \end{cases} \tag{2.4}$$

where the parameter θ is either estimated or set a priori. Common choices are the integers one and two, the latter following from the Newtonian gravity model. Another continuous weighting scheme is derived from the negative exponential function yielding

$$W_{ij} = \begin{cases} \exp(-\theta d_{ij}) & \text{if intercentroid distance } d_{ij} < d \, (d > 0, \theta > 0) \\ 0 & \text{otherwise} \end{cases} \tag{2.5}$$

where θ is a parameter that may be estimated, but is usually a priori chosen by the researcher. A popular choice is $\theta = 2$.

Evidently, a large number of spatial weights matrices can be derived for the same spatial layout. It is important to always keep in mind that the results of any spatial statistical analysis are conditional on the spatial weights matrix chosen. It is often good practice to check the sensitivity of the conclusions to the choice of the spatial weights matrix, unless there is a compelling reason on theoretical grounds to consider just a single one.

2.3 Global Measures and Tests for Spatial Autocorrelation

Spatial autocorrelation (association) is the correlation among observations of a single variable (*auto* meaning self) strictly attributable to the proximity of those observations in geographic space. This notion is best summarised by Tobler's first law which states that *"everything is related to everything else, but near things are more related than distant things"* (Tobler 1970, p. 234). Today, a number of measures of spatial autocorrelation are available (see Getis 2010 for a review).

Spatial autocorrelation measures deal with covariation or correlation between neighbouring observations of a variable. And thus compare two types of information: similarity of observations (value similarity) and similarity among locations (Griffith 2003). To simplify things, we will use the following notation

n number of areas in the sample,
i,j any two of the areal units,
z_i the value (observation) of the variable of interest for region i,
W_{ij} the similarity of i's and j's locations, with $W_{ii} = 0$ for all i,
M_{ij} the similarity of i's and j's observations of the variable.

Spatial autocorrelation (association) measures and tests may be differentiated by the scope or scale of analysis. Generally one distinguishes between global and local measures. Global implies that all elements in the W matrix are brought to bear on an assessment of spatial autocorrelation. That is, all spatial associations of areas are included in the calculation of spatial autocorrelation. This yields one value for spatial autocorrelation for any one spatial weights matrix. In contrast, local measures are focused. That is, they assess the spatial autocorrelation associated with one or a few particular areal units.

Global measures of spatial autocorrelation compare the set of value (observation) similarity M_{ij} with the set of spatial similarity W_{ij}, combining them into a single index of a cross-product, that is

$$\sum_{i=1}^{n}\sum_{j=1}^{n} M_{ij}\, W_{ij}. \tag{2.6}$$

In other words, the total obtained by multiplying every cell in the W matrix with its corresponding entry in the M matrix, and summing. Adjustments are made to each index to make it easy to interpret (see below).

Various ways have been suggested for measuring value similarity (association) M_{ij}, dependent upon the scaling of the variable. For nominal variables, the approach is to set M_{ij} to one if i and j take the same variable value, and zero otherwise. For ordinal variables, value similarity is generally based on comparing the ranks of i and j. For interval variables both the squared difference $(z_i - z_j)^2$ and the product $(z_i - \bar{z})(z_j - \bar{z})$ are commonly used, where $\bar{z}$ denotes the average of the z-values.

The two measures that have been most widely used for the case of areal units and interval scale variables are Moran's I and Geary's c statistics. Both indicate the degree of spatial association as summarised for the whole data set. Moran's I uses cross-products to measure value association, and Geary's c squared differences. Formally, Moran's I is given by the expression (see Cliff and Ord 1981, p. 17)

$$I = \frac{n}{W_o} \frac{\sum_{i=1}^{n} \sum_{j=1}^{n} W_{ij}(z_i - \bar{z})(z_j - \bar{z})}{\sum_{i=1}^{n} (z_i - \bar{z})^2} \tag{2.7}$$

with the normalising factor

$$W_o = \sum_{i=1}^{n} \sum_{j \neq i}^{n} W_{ij}. \tag{2.8}$$

For ease of interpretation the spatial weights W_{ij} may be in row-standardised form, though this is not necessary, and by convention $W_{ii} = 0$ for all i. Note that for a row-standardised W, $W_o = n$.

Geary's c is estimated as (Cliff and Ord 1981, p. 17)

$$c = \frac{(n-1)}{2\,W_o} \frac{\sum_{i=1}^{n} \sum_{j=1}^{n} W_{ij}(z_i - z_j)^2}{\sum_{i=1}^{n} (z_i - \bar{z})^2} \tag{2.9}$$

where W_o is given by Eq. (2.8). Neither of these statistics is constrained to lie in the $(-1, 1)$ range as in the case of conventional non-spatial product moment correlation. This is unlikely to present a practical problem for most real world data sets and reasonable W matrices (Bailey and Gatrell 1995, p. 270).

Spatial autocorrelation tests are decision rules based on statistics such as Moran's I and Geary's c to assess the extent to which the observed spatial arrangement of data values departs from the null hypothesis that space does not matter. This hypothesis implies that near-by areas do not affect one another such that there is independence and spatial randomness.

In contrast, under the alternative hypothesis of spatial autocorrelation (spatial association, spatial dependence), the interest renders on cases where large values are surrounded by other large values in near-by areas, or small values are surrounded by large values and vice versa. The former is referred to as *positive* spatial autocorrelation, and the latter as *negative* spatial autocorrelation. Positive spatial autocorrelation implies a spatial clustering of similar values (see Fig. 2.3a), while negative spatial autocorrelation implies a checkerboard pattern of values (see Fig. 2.3b).

Spatial autocorrelation is considered to be present when the spatial autocorrelation statistic computed for a particular pattern takes on a larger value, compared to what would be expected under the null hypothesis of no spatial

(a)   **(b)**

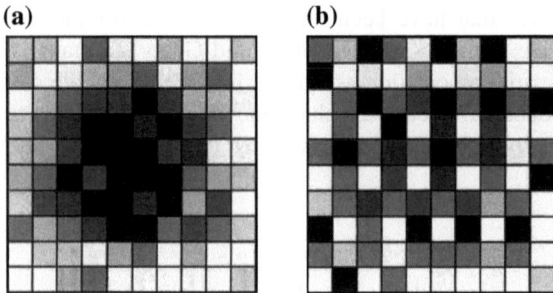

Fig. 2.3 Patterns of spatial autocorrelation on a regular grid: **a** positive spatial autocorrelation where cells with similar values (*gray tones*) are near-by; and **b** negative spatial autocorrelation where near-by cells have dissimilar values

association. What is viewed to be significantly larger depends on the distribution of the test statistic. We consider this question for the case of Moran's I statistic next.

In principle, there are two main approaches to testing observed I-values for significant departure from the hypothesis of zero spatial autocorrelation (Cliff and Ord 1981, p. 21). The first is the *random permutation test*. Under the randomisation assumption the observed value of I is assessed relative to the set of all possible values that could be obtained by randomly permuting the observations over the locations in the data set. Suppose we have n observations, z_i, relating to the a priori given areal units $i = 1, \ldots, n$.

Then $n!$ permutations are possible, each corresponds to a different arrangement of the n observations, z_i, over the areal units. One of these relates to the observed arrangement. The Moran I statistic can be computed for any of these $n!$ permutations. The resulting empirical distribution function provides the basis for a statement about the extremeness (or lack of extremeness) of the observed statistic, relative to the values computed under the null hypothesis (the randomly permuted values).

But computation of as many as $n!$ arrangements will be infeasible, even in the case of smaller n, since, for example, for $n = 10$ already 3,628,000 I-values would have to be calculated. But a close approximation to the permutation distribution can be obtained by using a Monte Carlo approach and simply sampling randomly from a reasonable number of the $n!$ possible permutations. Note that permutation re-orders the original data, whereas a Monte Carlo procedure generates "new" data of similar structure.

The other approach to testing observed I-values for significant departure from the hypothesis of zero spatial autocorrelation is based on an *approximate sampling distribution of I*. If there is a moderate number of areal units then an approximate result for the sampling distribution of I under certain assumptions may be utilised to develop a test. If it is assumed that the z_i are observations on random variables Z_i whose distribution is normal, then I has a sampling distribution that is appropriately normal with the moments

$$E(I) = -\frac{1}{(n-1)} \qquad (2.10)$$

$$var\,(I) = \frac{n^2(n-1)\,W_1 - n\,(n-1)\,W_2 - 2\,W_o^2}{(n+1)\,(n-1)^2\,W_o^2} \qquad (2.11)$$

where

$$W_o = \sum_{i=1}^{n} \sum_{j \neq i}^{n} W_{ij} \qquad (2.8)$$

$$W_1 = \frac{1}{2} \sum_{i=1}^{n} \sum_{j \neq i}^{n} \left(W_{ij} + W_{ji}\right)^2 \qquad (2.12)$$

$$W_2 = \sum_{k=1}^{n} \left(\sum_{j=1}^{n} W_{kj} + \sum_{i=1}^{n} W_{ik}\right)^2 . \qquad (2.13)$$

Hence, we can test the observed value of I against the percentage points of the appropriate sampling distribution. An "extreme" observed value of I indicates significant spatial autocorrelation. A value of Moran's I that exceeds its expected value of $-1/(n-1)$ points to *positive* spatial autocorrelation, while a value of Moran's I that is below the expectation indicates *negative* spatial autocorrelation (Bailey and Gatrell 1995, pp. 281–282).

Note that the hypothesis involved in each of the above two tests is somewhat different. The randomisation test embodies the assumption that no values of z_i other than those observed are realisable. In other words the data are treated as a population and the question analysed is how the data values are arranged spatially. Hence, the test is a test of patterns in the observations relative to the set of all possible patterns in the *given* observations. The approximate sampling distribution test makes the assumption that the observations z_i are observations on (normal) random variables Z_i. That is, they are one realisation of a random process and other possible realisations can occur. The test is, thus, one of spatial autocorrelation, providing the distribution of the random variables Z_i can be assumed to be normal (Bailey and Gatrell 1995, pp. 280–282; Fortin and Dale 2009).

Care is necessary in applying the above formal tests of spatial autocorrelation when I has been computed from residuals that arise from a regression (see Chap. 3). The problem arises because if Q parameters (regression coefficients β_q, $q = 1, \ldots, Q$) have been estimated in the regression, then the observed residuals are subject to Q linear constraints. That is, the observed residuals will be automatically spatially autocorrelated to some extent, and consequently the above testing procedure for Moran's I will not be valid. If $Q \ll n$, however, then it might be justified in ignoring this. If not, then strictly one should use adjustments to the mean and variance of the approximate sampling distribution of I. We do not go into details here, but refer the reader to Chap. 3 and to the literature cited therein which also covers tests for spatial autocorrelation at spatial lags.

2.4 Local Measures and Tests for Spatial Autocorrelation

With the advent of large data sets characteristic of GISystems, it has become clear that the need to assess spatial autocorrelation globally may be of only marginal interest. During the past two decades, a number of statistics, called local statistics, have been developed. These provide for each observation of a variable an indication of the extent of significant spatial clustering of similar values around that observation. Hence, they are well suited to identify the existence of *hot spots* (local clusters of high values) or *cold spots* (local clusters of low values), and are appropriate to identify distances beyond which no discernible association exists.

Let us assume that each area i $(i = 1, \ldots, n)$ has associated with it a value z_i that represents an observation upon the random variable Z_i. Typically, it is assumed that the Z_i have identical marginal distributions. If they are independent, we say that there is no spatial structure. Independence implies the absence of spatial autocorrelation. But the converse is not necessarily true. Nevertheless, tests for spatial autocorrelation are characteristically viewed as appropriate assessment of spatial dependence (association). Usually, if spatial autocorrelation exists, it will be exhibited by similarities among neighbouring areas, although negative patterns of spatial association are also possible (Ord and Getis 1995, p. 287).

The basis for local tests for and measures of spatial autocorrelation comes from the cross-product statistic

$$\sum_{j=1}^{n} M_{ij} W_{ij} \qquad (2.14)$$

that allows for spatial autocorrelative comparisons for a given observation (areal unit) $i = 1, \ldots, n$ where M_{ij} and W_{ij} are defined as in the previous section. We briefly describe four local statistics: the Getis and Ord local statistics G_i and G_i^*, and the local versions of Moran's I and Geary's c. Let us begin with the local statistics suggested by Getis and Ord (1992). The statistics are computed by defining a set of neighbours for each area i as those observations that fall within a critical distance d from i where each $i = 1, \ldots, n$ is identified with a point (centroid). This can be formally expressed in a set of symmetric binary weights matrices $W(d)$, with elements $W_{ij}(d)$ indexed by distance d. For each distance d, the elements $W_{ij}(d)$ of the corresponding weights matrix $W(d)$ equal one if i and j are within a distance from each other, and zero otherwise. Clearly, for different distance measures, a different set of neighbours will be found.

The G_i and G_i^* statistics measure the degree of local association for each observation i in a data set containing n observations. They consist of the ratio of the sum of values in neighbouring areas, defined by a given distance band, to the sum over all observations (excluding the value in area i for the G_i statistic, but including it for the G_i^* statistic). These statistics may be computed for many different distance bands. Formally, the G_i measure for observation (area) i can be expressed as

$$G_i(d) = \frac{\sum_{j \neq i}^{n} W_{ij}(d)\, z_j}{\sum_{j \neq i}^{n} z_j} \tag{2.15}$$

with the summation in j exclusive of i. The G_i^* measure is given by

$$G_i^*(d) = \frac{\sum_{j=1}^{n} W_{ij}(d)\, z_j}{\sum_{j=1}^{n} z_j} \tag{2.16}$$

except that the summation in j is now inclusive of i. The G_i statistic can be interpreted as a measure of clustering of like values around a particular observation i, irrespective of the value in that area, while the G_i^* statistic includes the value within the measure of clustering. A positive value indicates clustering of high values and a negative value indicates a cluster of low values. It is interesting to note that G_i^* is mathematically associated with global Moran's $I(d)$ so that Moran's I may be interpreted as a weighted average of local statistics (Getis and Ord 1992). A slightly different form of the G-statistic was suggested by Ord and Getis (1995) where the distributional characteristics are discussed in detail (see also Getis 2010).

Getis and Ord (1992), and Ord and Getis (1995) provide the expected values and variances of the two statistics. Their distribution is normal if the underlying distribution of the observations is normal. But if the distribution is skewed, the test only approaches normality as the critical distance d increases, and does so more slowly for boundary areas where there are fewer neighbours. In other words, under these circumstances normality of the test statistic can only be guaranteed when the number of j neighbouring areas is large. When n is relatively small, as few as eight neighbours could be used without serious inferential errors unless the underlying distribution is very skewed (Getis and Ord 1996). Hot spots identified by these statistics can be interpreted as clusters or indications of spatial non-stationarity.

Local indicators of spatial association (LISA) *statistics* were derived by Anselin (1995), with the motivation to decompose global spatial autocorrelation statistics, such as Moran's I and Geary's c, into the contribution of each individual observation $i = 1, \ldots, n$. The local Moran statistic I_i for observation (area) $i = 1, \ldots, n$ is defined (Anselin 1995) as

$$I_i = (z_i - \bar{z}) \sum_{j \in J_i}^{n} W_{ij} (z_j - \bar{z})^2 \tag{2.17}$$

where J_i denotes the neighbourhood set of area i, and the summation in j runs only over those areas belonging to J_i, $\bar{z}$ denotes the average of these neighbouring observations.

It is evident that the sum of I_i for all observations i

$$\sum_{i=1}^{n} I_i = \sum_{i=1}^{n} (z_i - \bar{z}) \sum_{j=J_i}^{n} W_{ij}(z_j - \bar{z}) \tag{2.18}$$

is proportional to the global Moran statistic I given by Eqs. (2.7) and (2.8).

The moments for I_i under the null hypothesis of no spatial association can be derived using the principles outlined in Cliff and Ord (1981, pp. 42–46). For example, for a randomisation hypothesis, the expected value is found as

$$E[I_i] = -\frac{1}{(n-1)} \tilde{W}_i \tag{2.19}$$

and the variance turns out to be

$$var[I_i] = \frac{1}{(n-1)} W_{i(2)}(n - b_2) + \frac{2}{(n-1)(n-2)} W_{i(kh)}(2b_2 - n) - \frac{1}{(n-1)^2} \tilde{W}_i^2 \tag{2.20}$$

where

$$W_{i(2)} = \sum_{j \neq i}^{n} W_{ij}^2 \tag{2.21}$$

$$2W_{i(kh)} = \sum_{k \neq i}^{n} \sum_{h \neq i}^{n} W_{ik} W_{ih} \tag{2.22}$$

$$\tilde{W}_i = \sum_{j=1}^{n} W_{ij} \tag{2.23}$$

with $b_2 = m_4 m_2^{-2}$, $m_2 = \Sigma_i(z_i - z)^2 n^{-1}$ as the second moment, and $m_4 = \Sigma_i(z_i - z)^4 n^{-1}$ as the fourth moment. A test for significant local spatial association may be based on these moments, although the exact distribution of such a statistic is still unknown (Anselin 1995, p. 99).

Alternatively, a conditional random permutation test can be used to yield so-called *pseudo significance levels*. The randomisation is conditional in the sense that the value z_i associated with area i is hold fixed in the permutation, and the remaining values are randomly permuted over the areas. For each of these resampled data sets, the value of the local Moran I_i can be computed. The resulting empirical distribution function provides the basis for a statement about the extremeness or lack of extremeness of the observed statistic I_i, relative—and conditional on—the I_i-values computed under the null hypothesis.

A complicating factor in the assessment of significance is that the statistics for individual locations (areas) will tend to be correlated whenever the neighbourhood sets J_i and J_k of two areas i and k contain common elements. Due to this

correlation, and the associated problem of multiple comparisons, the usual interpretation of significance will be flawed. Furthermore, it is impossible to derive the
exact marginal distribution of each statistic, and the significance levels have to be
approximated by Bonferroni inequalities or following the approach suggested by
Sidák (1967). This means—as pointed out by Anselin (1995, p. 96)—that when the
overall significance associated with the multiple comparisons (correlated tests) is
set to α, and there are m comparisons, then the individual significance α_i should be
set to either α/m (Bonferroni) or $1 - (1 - \alpha)^{1/m}$ (Sidák). Note that the use of
Bonferroni bounds may be too conservative for local indicators of association.
If, for example, $m = n$, then an overall significance of $\alpha = 0.05$ would imply
individual levels of $\alpha_i = 0.0005$ in a data set with one hundred observations,
possibly revealing only very few if any significant areas. But since the correlation
between individual statistics is due to the common elements in the neighbourhood
sets, only for a small number of areas k will the statistics actually be correlated
with an individual I_i (Anselin 1995, p. 96).

Using the same notation as before, a local Geary statistic c_i for each observation
$i\,(i = 1, \ldots, n)$ may be defined as

$$c_i = \sum_{j \in J_i}^{n} W_{ij}(z_i - z_j)^2 \tag{2.24}$$

where J_i denotes the neighbourhood set of area i. The c_i statistic is interpreted in
the same way as the local Moran. The summation of the c_i over all observations
yields

$$\sum_{i=1}^{n} c_i = \sum_{i=1}^{n} \sum_{j \in J_i}^{n} W_{ij}(z_i - z_j)^2 \tag{2.25}$$

that is evidently proportional to the global Geary c statistic given by Eq. (2.9).

These LISA statistics, I_i and c_i, serve two purposes. On the one hand, they may
be viewed as indicators of local pockets of non-stationarity, or hot spots, similar
to the G_i and G_i^* statistics. On the other hand, they may be used to assess
the influence of individual locations (observations) on the magnitude of the corresponding global spatial autocorrelation statistic, Moran's I and Geary's c.

Chapter 3
Modelling Area Data

Abstract Exploratory spatial data analysis is often a preliminary step to more formal modelling approaches that seek to establish relationships between the observations of a variable and the observations of other variables, recorded for each areal unit. The focus in this chapter is on spatial regression models in a simple cross-sectional setting, leaving out of consideration the analysis of panel data. We, moreover, assume that the data concerned can be taken to be approximately normally distributed. This assumption is—to varying degrees—involved in most of the spatial regression techniques that we will consider. Note that the assumption of normality is not tenable if the variable of interest is a count or a proportion. In these cases we would expect models for such data to involve probability distributions such as the Poisson or binomial. The chapter consists of five sections, starting with a treatment of the specification of spatial dependence in a regression model. Next, specification tests are considered to detect the presence of spatial dependence. This is followed by a review of the spatial Durbin model (SDM) that nests many of the models widely used in the literature, and by a discussion of spatial regression model estimation based on the maximum likelihood (ML) principle. The chapter closes with some remarks on model parameter interpretation, an issue that had been largely neglected so far. Readers interested in implementing the models, methods and techniques discussed in this chapter find useful MATLAB code which is publicly available at spatial-econometrics.com, LeSage's spatial econometrics toolbox (downloadable from http://www.spatial-econometrics.com/), see Liu and LeSage (2010) Journal of Geographical Systems 12(1):69–87 for a brief description. Another useful open software is the spdep package of the R project (downloadable from http://cran.r-project.org).

Keywords Area data · Spatial regression models · Spatial lag models · Spatial error models · Higher order models · Spatial Durbin model · Specification search · Tests for spatial dependence · Maximum likelihood estimation · Model parameter interpretation

M. M. Fischer and J. Wang, *Spatial Data Analysis*,
SpringerBriefs in Regional Science, DOI: 10.1007/978-3-642-21720-3_3,

3.1 Spatial Regression Models

Starting point is the linear regression model, where for each observation (area) i, with $i = 1, \ldots, n$, the following relationship holds

$$y_i = \sum_{q=1}^{Q} X_{iq}\beta_q + \varepsilon_i \qquad (3.1)$$

where y_i is an observation on the dependent variable, X_{iq} is an observation on an explanatory variable, with $q = 1, \ldots, Q$ (including a constant, or one), β_q the matching regression coefficient, and ε_i the error term.

In the classical regression specification, the error terms have mean zero, that is, $E[\varepsilon_i] = 0$ for all i, and they are identically and independently distributed (*iid*). Hence, their variance is constant, $\text{var}[\varepsilon_i] = \sigma^2$ for all i, and they are uncorrelated, $E[\varepsilon_i \varepsilon_j] = E[\varepsilon_i]E[\varepsilon_j] = 0$ for $i \neq j$.

In matrix notation this regression model may be written as

$$y = X\beta + \varepsilon \qquad (3.2)$$

where the n observations on the dependent variable are stacked in an n-by-1 vector y, the observations on the explanatory variables in an n-by-Q matrix X with the associated Q-by-1 parameter vector β, and the random error terms in an n-by-1 vector ε. $E[\varepsilon] = 0$ where 0 is an n-by-1 vector of zeros, and $E[\varepsilon \varepsilon'] = \sigma^2 I$ with I denoting the n-by-n identity matrix.

The assumption of independent observations greatly simplifies the model, but in the context of area data this simplification is very unlikely to be appropriate, because of the possibility of spatial dependence between the error terms. If the regressors, residuals or the dependent variable are spatially dependent, the model suffers from a misspecification problem and the results of the model are biased or inconsistent.

Spatial dependence reflects a situation where values observed in one areal unit, depend on the values of neighbouring observations at near-by areas. Spatial dependence may be introduced into a model of type (3.2) in two major ways: one is referred to as spatial lag dependence, and the other as spatial error dependence (Anselin 1988b). The former pertains to spatial correlation in the dependent variable, while the latter refers to the error term. Hence, it has become convenient to distinguish between spatial lag and spatial error model specifications.

Spatial dependence can also be introduced in the regressor variables, leading to so-called cross-regressive models (Florax and Folmer 1992), also termed spatially lagged X (or SLX) models (LeSage and Pace 2009). But—in contrast to the spatial lag and spatial error models—they do not require the application of special estimation procedures. Thus, they will not be further considered in this chapter.

Spatial lag models account for spatial correlation (dependence) in the dependent variable. Such specifications are typically motivated by theoretical arguments that emphasise the importance of neighbourhood effects, or spatial externalities that cross the borders of the areal units and show up in the dependent variable. This kind of spatial autocorrelation is substantive on the ground that it has a meaningful interpretation.

In contrast, spatial error models account for spatial dependence in the error term. Spatial error dependence may arise, for example, from unobservable latent variables that are spatially correlated. It may also arise from area boundaries that do not accurately reflect neighbourhoods which give rise to the variables collected for the analysis. Spatial autocorrelation arising for these reasons is considered to be nuisance.

Spatial Lag Models Spatial lag models are extensions of regression models of type (3.1). They allow observations of the dependent variable y in area i ($i = 1, \ldots, n$) to depend on observations in neighbouring areas $j \neq i$. The basic spatial lag model, the so-called first order *spatial autoregressive* (SAR) model, takes the form

$$y_i = \rho \sum_{j=1}^{n} W_{ij} y_j + \sum_{q=1}^{Q} X_{iq} \beta_q + \varepsilon_i \tag{3.3}$$

where the error term, ε_i, is *iid*. W_{ij} is the (i, j)th element of the n-by-n spatial weights matrix W (see Sect. 2.2). Recall that W has non-zero elements W_{ij} in each row i for those columns j that are neighbours of area i. By convention, $W_{ii} = 0$ for all i. All these values are exogenous. We assume that W is row-stochastic so that the matrix W has a principal eigenvalue of one. The term row-stochastic refers to a non-negative matrix having row sums normalised so they equal one.

The scalar ρ in Eq. (3.3) is a parameter (to be estimated) that will determine the strength of the spatial autoregressive relation between y_i and $\Sigma_j W_{ij} y_j$, a linear combination of spatially related observations based on non-zero elements in the ith row of W. The domain of ρ is defined by the interval $(w_{min}^{-1}, w_{max}^{-1})$, where w_{min} and w_{max} represent the minimum and maximum eigenvalues of the matrix W. For the case of a row-normalised weights matrix, $-1 \leq w_{min} < 0$, $w_{max} = 1$ so that ρ ranges from negative values to unity. In cases where positive spatial dependence is almost certain, restriction of ρ to the $[0, 1)$ interval simplifies computation. It should be clear that if $\rho = 0$, we have a conventional regression model of type (3.1) so that interest focuses on the statistical significance of the coefficient estimate for ρ.

In matrix notation, model (3.3) may be written as

$$y = \rho W y + X \beta + \varepsilon. \tag{3.4}$$

With a row-standardised spatial weights matrix W (that is, the weights are standardised such that $\Sigma_j W_{ij} = 1$ for all i), this amounts to including the average of the neighbours as an additional variable into the regression specification. This variable, Wy, is referred to as a spatially lagged dependent variable. For example, in a model for growth rates of European regions, this would add the average of the growth rates in the neighbouring locations as an explanatory variable.The model given by Eq. (3.4) is a structural model. Its reduced form, that is, the solution of the model for y is

$$y = (I - \rho W)^{-1}(X\beta + \varepsilon) \qquad (3.5)$$

so that the expected value of y is

$$E[y] = (I - \rho W)^{-1}X\beta \qquad (3.6)$$

since the errors all have mean zero. The inverse matrix term is called *spatial multiplier*, and indicates that the expected value of each observation y_i will depend on a linear combination of X-values taken by neighbouring observations, scaled by the dependence parameter ρ.

Spatial Error Models Another form of spatial dependence occurs when the dependence works through the error process, in that the errors from different areas may display spatial covariance. The most common specification is a spatial autoregressive process of first order, as given by

$$\varepsilon_i = \lambda \sum_{j=1}^{n} W_{ij}\varepsilon_j + u_i \qquad (3.7)$$

where λ is the autoregressive parameter, and u_i a random error term, typically assumed to be *iid*. In matrix notation Eq. (3.7) may be reformulated as

$$\varepsilon = \lambda W \varepsilon + u. \qquad (3.8)$$

Assuming $|\lambda| < 1$ and solving Eq. (3.8) for ε yields

$$\varepsilon = (I - \lambda W)^{-1} u. \qquad (3.9)$$

Inserting Eq. (3.9) into the standard regression model (3.2) yields the spatial error model

$$y = X\beta + (I - \lambda W)^{-1} u \qquad (3.10)$$

with $E[uu'] = \sigma^2 I$ so that the complete error variance–covariance matrix follows as

$$E[\varepsilon\,\varepsilon'] = \sigma^2 (I - \lambda W)^{-1}(I - \lambda W')^{-1}. \qquad (3.11)$$

The spatial error model (SEM) may be viewed as a combination of a standard regression model with a spatial autoregressive model in the error term ε, and hence has an expectation equal to that of the standard regression model. In large samples, point estimates for the parameters β from the SEM model and conventional regression will be the same, but in small samples there may be an efficiency gain from correctly modelling spatial dependence in the error terms. Note that in contrast spatial lag models that contain spatial lag terms Wy on the right-hand side of the equation generate expectations that differ from those of the standard regression model.

Higher Order Models In addition to the basic spatial lag and spatial error models described above, higher order models can be specified as well, by including two or more weight matrices. Using multiple weight matrices provides a straightforward generalisation of the SAR and SEM models. For example, Anselin (1988b, pp. 34–36) uses two spatial weights matrices W_1 and W_2 to combine the basic spatial lag and error models so that

$$y = \rho W_1 y + X\beta + \varepsilon \qquad (3.12)$$

$$\varepsilon = \lambda W_2 \varepsilon + u \qquad (3.13)$$

$$u \sim \mathcal{N}(0, \sigma_u^2 I) \qquad (3.14)$$

where W_1 and W_2 represent n-by-n non-negative spatial weights matrices (not necessarily distinct from each other) with zeros on the main diagonal. The parameters to be estimated are β, ρ, λ and σ_u^2. Setting the parameter $\rho = 0$ eliminates the spatially lagged variable $W_1 y$, generating the basic spatial error model given by Eq. (3.10). The case where $\lambda = 0$ eliminates the spatially lagged disturbance term yields the basic spatial lag model given by Eq. (3.4).

3.2 Tests for Spatial Dependence

The standard approach towards detecting the presence of spatial dependence in a regression model is to apply diagnostic tests. The best known test statistic against spatial autocorrelation is Moran's I statistic for spatial autocorrelation applied to the regression residuals (see Cliff and Ord 1972, 1973, see also Sect. 2.3):

$$I = \frac{n}{W_0} \frac{e'We}{e'e} \qquad (3.15)$$

$$W_0 = \sum_{i=1}^{n} \sum_{j \neq i}^{n} W_{ij} \qquad (3.16)$$

where e is an n-by-1 vector of OLS residuals $y - X\hat{\beta}$, $e'e$ is the sum of squared residuals, and W_0, equal to the sum of W_{ij} over i and j, is a normalising factor. Note that the correcting factor n/W_0 is not needed if the spatial weights matrix W is row-standardised. In practice, inference—by means of Moran's I test—is based on a normal approximation, using a standardised value, obtained by subtracting the mean under the null and dividing by the square root of the variance.

As already pointed out in Sect. 2.3, care needs to be taken when applying this formal test of spatial dependence to residuals. The problem arises because if Q regression coefficients have been estimated, then the observed residuals are automatically subject to Q linear constraints. That is, the observed residuals will be correlated to some extent, and hence the testing procedure for Moran's I will not be valid. If $Q \ll n$, however, then it might be justified in ignoring this. If not, then strictly one should use adjustments to the mean and variance of the approximate sampling distribution of I.

An alternative, more focused test for spatial error dependence is based on the Lagrange multiplier (LM) principle, suggested by Burridge (1980). It is similar in expression to Moran's I and is also computed from the OLS residuals. But a normalisation factor in terms of matrix traces is needed to achieve an asymptotic chi-square distribution (with one degree of freedom) under the null hypothesis of no spatial dependence $(H_0 : \lambda = 0)$. The LM error statistic is given by

$$LM(\text{error}) = \left(\frac{e'We}{e'e\,n^{-1}} \right)^2 \frac{1}{tr[W'W + W^2]} \qquad (3.17)$$

where tr stands for the trace operator (the sum of the diagonal elements of a matrix), and $(e'e\,n^{-1})$ represents the error variance. Except for the scaling factor $tr[W'W + W^2]^{-1}$, this statistic is essentially the square of Moran's I.

A test for substantive spatial dependence (that is, an omitted spatial lag) can also be based on the Lagrange multiplier principle (see Anselin 1988b). Its form is slightly more complex, but again requires only the results of an OLS regression. The test takes the form

$$LM(\text{lag}) = \left(\frac{e'Wy}{e'e\,n^{-1}} \right)^2 \frac{1}{H} \qquad (3.18)$$

with

$$H = \{(WX\hat{\beta})'[I - X(X'X)^{-1}X'](WX\hat{\beta})\hat{\sigma}^{-2}\} + tr(W'W + W^2) \qquad (3.19)$$

where $\hat{\beta}$ and $\hat{\sigma}^2$ denote OLS estimates, Wy is the spatial lag and $WX\hat{\beta}$ is a spatial lag for the predicted values $X\hat{\beta}$, and $[I - X(X'X)^{-1}X']$ is a familiar projection matrix. The LM(lag) test is also distributed as chi-square with one degree of freedom under the null hypothesis of no spatial dependence $(H_0 : \rho = 0)$.

Specification Search For the simple case of choosing between a spatial lag or spatial error alternative, there is considerable evidence that the proper alternative is most likely the one with the largest significant LM test statistic value (Anselin and Rey 1991). This was later refined in light of the robust form of the two LM statistics in Anselin et al. (1996) accounting for the fact that in the presence of spatial lag (error) dependence, the LM test against error (lag) dependence becomes biased.

Florax and Folmer (1992) suggest a sequential testing procedure to discern whether a model based on the restrictions $\rho = 0$ versus $\lambda = 0$, versus both ρ and λ different from zero should be selected. Of course, this approach complicates inference concerning the parameters of the final model specification due to the pre-test issue.

Florax et al. (2003) consider Hendry's "general to specific approach" to model specification versus a forward stepwise strategy. While the "general to specific" approach tests sequential restrictions placed on the most general model

(3.12)–(3.14) that includes both spatial lag and spatial error dependence, the stepwise strategy considers sequential expansions of the model. Starting with regression model (3.2), expansion of the model proceeds to add spatial lag terms, conditional upon the results of misspecification tests. They conclude that the Hendry approach is inferior in its ability to detect that true data generating process.

3.3 The Spatial Durbin Model

The spatial Durbin model (SDM) is the SAR model (3.4) augmented by spatially lagged explanatory variables

$$y = \rho W y + X \beta + W \bar{X} \gamma + \varepsilon \tag{3.20}$$

where $\bar{X}$ is the n-by-$(Q-1)$ non-constant explanatory variable matrix. The model may be rewritten in reduced form as

$$y = (I - \rho W)^{-1} (X \beta + W \bar{X} \gamma + \varepsilon) \tag{3.21}$$

with

$$\varepsilon = \mathcal{N}(0, \sigma^2 I) \tag{3.22}$$

where γ is a $(Q-1)$-by-1 vector of parameters that measure the marginal impact of the explanatory variables from neighbouring observations (areas) on the dependent variable y. Multiplying $\bar{X}$ by W (that is, $W\bar{X}$) produces spatially lagged explanatory variables that reflect an average of neighbouring observations. If W is sparse (having a large proportion of zeros), operations such as $W\bar{X}$ require little time.

By defining $Z = [X \ W\bar{X}]$ and $\delta = [\beta \ \gamma]'$ this model can be written as a SAR model leading to

$$y = \rho W y + Z \delta + \varepsilon \tag{3.23}$$

or

$$y = (I - \rho W)^{-1} Z \delta + (I - \rho W)^{-1} \varepsilon. \tag{3.24}$$

One motivation for use of the SDM model rests on the plausibility of a conjunction of two circumstances that seem likely to arise in applied spatial regression modelling of area data samples. One of these is spatial dependence in the disturbances of an OLS regression model. The second circumstance is the existence of an omitted explanatory variable that exhibits non-zero covariance with a variable included in the model, and omitted variables are likely when dealing with areal data samples (LeSage and Fischer 2008).

In addition, the spatial Durbin model occupies an interesting position in the field of spatial regression analysis because it nests many of the models widely used in the literature (see LeSage and Pace 2009):

(i) imposing the restriction $\gamma = 0$ leads to the spatial autoregressive model (3.4) that includes a spatial lag of the dependent variable, but excludes the influence of the spatially lagged explanatory variables,

(ii) the so-called common factor parameter restriction $\gamma = -\rho\,\bar{\beta}$ yields the spatial error regression model specification (3.10) which assumes that externalities across areas are mostly a nuisance spatial dependence problem caused by the spatial transmission of random shocks (Note that $\bar{\beta}$ denotes the $(Q{-}1)$-by-1 vector of parameters that measure the marginal impact of the non-constant explanatory variables on the dependent variable. $\beta = (\beta_0, \bar{\beta})'$ where β_0 is the constant term parameter),

(iii) the restriction $\rho = 0$ results in a least squares spatially lagged X regression model that assumes independence between observations of the dependent variable, but includes characteristics from neighbouring areas, in the form of spatially lagged explanatory variables,

(iv) finally, imposing the restriction $\rho = 0$ and $\gamma = 0$ yields the standard least squares regression model given by Eq. (3.2).

Hence, the SDM model suggests a general-to-simple model selection rule. Testing, whether the restrictions hold or not, implies not much effort. Of particular importance are common factor tests that discriminate between the unrestricted SDM and the SEM specifications, or in other words between substantive and residual dependence in the analysis. The likelihood ratio test proposed by Burridge (1980) is the most popular test in this context (see LeSage and Pace (2009) for details, Mur and Angulo (2006) for alternative tests and a comparison based on Monte Carlo evidence).

Finally, it should be noted that the spatial Durbin model (3.20) can be generalised to

$$y = \rho W_1 y + X\beta + W_1 \bar{X}\gamma + \varepsilon \qquad (3.25)$$

$$\varepsilon = \lambda W_2 \varepsilon + u \qquad (3.26)$$

$$u \sim \mathcal{N}(0, \sigma_u^2 I), \qquad (3.27)$$

where the n-by-n spatial weights matrices W_1 and W_2 can be the same or distinct. For details on this model generalisation see LeSage and Pace (2009, pp. 52–54).

3.4 Estimation of Spatial Regression Models

Estimation of spatial regression models is typically carried out by means of a maximum likelihood (ML) approach, in which the probability of the joint distribution (likelihood) of all observations is maximised with respect to a number of relevant parameters. Maximum likelihood estimation has desirable asymptotic theoretical properties such as consistency, efficiency and asymptotic normality, and is also thought to be robust for small departures from the normality assumption (LeSage and Pace 2004, pp. 10–11).

The estimation problems associated with spatial regression models are different for the spatial lag and the spatial error cases. We start the discussion by focusing on the SAR (and SDM) model presented in Eq. (3.23).

Given $\varepsilon \sim \mathcal{N}(0, \sigma^2 I)$, the log (more precisely the logarithm naturalis) of the likelihood for the *SAR model* given by Eq. (3.23) takes the form in Eq. (3.28) (Anselin 1988b, p. 63)

$$
\ln L(\rho, \delta, \sigma^2) = -\frac{n}{2}\ln 2\pi - \frac{n}{2}\ln \sigma^2 + \ln |A|
$$
$$
- \frac{1}{2\sigma^2}(Ay - Z\delta)'(Ay - Z\delta) \tag{3.28}
$$

where n is the number of observations, $|.|$ stands for the determinant of a matrix, and for notational simplicity, the expression $I - \rho W$ is replaced by A. The parameters with respect to which this likelihood has to be maximised are ρ, δ and σ^2.

The minimisation of the last term in Eq. (3.28) corresponds to ordinary least squares (OLS), but since this ignores the log-Jacobian term $\ln |I - \rho W|$, OLS is not a consistent estimator in this model. There is no satisfactory two-step procedure and estimators for the parameters have to be obtained from an explicit maximisation of the likelihood (Anselin 2003b).

But it turns out that the estimates for the regressive coefficients δ, conditional upon the value for ρ, can be found as

$$
\delta = \delta_O - \rho \delta_L \tag{3.29}
$$

where δ_O and δ_L are OLS regression coefficients in a regression of Z on y and Wy, respectively. In a similar way the error variance σ^2 can be estimated as

$$
\sigma^2 = (e_O - \rho e_L)'(e_O - \rho e_L)\frac{1}{n} \tag{3.30}
$$

where e_O and e_L are the residual vectors in the regressions for δ_O and δ_L. That is, $e_O = y - Z\delta_O$ and $e_L = Wy - Z\delta_L$, where $\delta_O = (Z'Z)^{-1}Z'y$ and $\delta_L = (Z'Z)^{-1}Z'Wy$.

Substitution of (3.29) and (3.30) into the log-likelihood function (3.28) gives the scalar concentrated log-likelihood function value

$$
\ln L_{con}(\rho) = \kappa + \ln |I - \rho W| - \frac{n}{2}\ln\big[(e_O - \rho e_L)'(e_O - \rho e_L)\big] \tag{3.31}
$$

where κ is a constant that does not depend on ρ. The motivation for optimising the concentrated log-likelihood is that this simplifies the optimising problem by reducing a multivariate optimisation problem to a univariate one. Maximising the concentrated log-likelihood function with respect to ρ yields ρ^* that is equal to the maximum likelihood estimate $(\hat{\rho}_{ML} = \rho^*)$. Note that it is well-known that maximum likelihood often has a downward bias in estimation of ρ in small samples.

The computationally difficult aspect of this optimisation problem for models with a large number of observations is the need to compute the log-determinant of the n-by-n matrix $(I - \rho W)$. In response to this computational challenge there are at least two strategies. *First*, the use of alternative estimators can solve this problem. Examples include the instrumental variable (IV) approach (Anselin 1988b, pp. 81–90) and the instrumental variable (IV)/generalised moments (GM) approach (Kelejian and Prucha 1998, 1999). These alternative estimation methods, however, suffer from several drawbacks. One is that they can produce ρ-estimates that fall outside the interval defined by the eigenvalue bounds arising from the spatial weights matrix W. Moreover, inferential procedures for these methods can be sensitive to implementation issues such as the interaction between the choice of instruments and model specification which are not always obvious to the practitioner (LeSage and Pace 2010).

A *second* strategy is to directly attack the computational difficulties confronting ML estimation. The Taylor series approach of Martin (1993), the eigenvalue based approach of Griffith and Sone (1995), the direct sparse matrix approach of Pace and Barry (1997), the characteristic polynomial approach of Smirnov and Anselin (2001), and the sampling approach of Pace and LeSage (2009) are examples of this strategy. A review of most of the approximations to the log-determinant can be found in LeSage and Pace (LeSage and Pace 2009, Chap. 4). Improvements in computing technology in combination with these approaches suggest that very large problems can be handled today, using the ML estimation approach.

Inference regarding parameters for the models is frequently based on estimates of the variance–covariance matrix. In problems where the sample size is small, an asymptotic variance matrix based on the Fisher information matrix for parameters $\eta = (\rho, \delta, \sigma^2)$ can be used to provide measures of dispersion for these parameters. Anselin (1988b) provides the analytical expressions needed to construct this information matrix, but evaluating these expressions may be computationally difficult when dealing with large scale problems involving thousands of observations (LeSage and Pace 2004, p. 13).

Let us turn next to the spatial error model presented in Eq. (3.10) that represents another member of the family of regression models that can be derived from Eqs. (3.12)–(3.14). Assuming normality for the error terms, and using the concept of a Jacobian for this model as well, the log-likelihood for the SEM model can be obtained as

$$\ln L(\lambda, \beta, \sigma^2) = -\frac{n}{2}\ln 2\pi - \frac{n}{2}\ln \sigma^2 + \ln |I - \lambda W|$$

$$-\frac{1}{2\sigma^2}(y - X\beta)'(I - \lambda W)'(I - \lambda W)(y - X\beta). \quad (3.32)$$

A closer inspection of the last term in Eq. (3.32) reveals that—conditional upon a given λ—a maximisation of the log-likelihood is equivalent to the minimisation of the sum of squared residuals in a regression of a spatially filtered dependent variable $y^* = y - \lambda Wy$ on a set of spatially filtered explanatory variables $X^* = X - \lambda WX$. The first order conditions for $\hat{\beta}_{ML}$ indeed generate the familiar generalised least squares estimator (Anselin 2003b)

$$\hat{\beta}_{ML} = [X'(I - \lambda W)'(I - \lambda W)X]^{-1}X'(I - \lambda W)'(I - \lambda W)y \qquad (3.33)$$

and, similarly, the ML estimator for σ^2 as

$$\hat{\sigma}^2_{ML} = (e - \lambda We)'(e - \lambda We)\frac{1}{n} \qquad (3.34)$$

where $e = y - X\hat{\beta}_{ML}$. A consistent estimate for λ cannot be obtained from a simple auxiliary regression, but the first order conditions must be solved explicitly by numerical means. For technical details, see Anselin (1988b, Chap. 6), or LeSage and Pace (2009, Chap. 3). As for the spatial lag model, asymptotic inference can be based on the inverse of the information matrix (see Anselin 1988b, Chap. 6, for details).

3.5 Model Parameter Interpretation

Simultaneous feedback is a feature of spatial regression models that comes from dependence relations embodied in the spatial lag term Wy. This leads to feedback effects from changes in explanatory variables in an area that neighbours i, say area j, that will impact the dependent variable for observation (area) i. Consequently, interpretation of parameters of spatial regression models that contain a spatial lag Wy becomes more complicated (see, for example, Kim et al. 2003; Anselin and LeGallo 2006; Kelejian et al. 2006; LeSage and Fischer 2008).

To see how these feedback effects work, we follow LeSage and Pace (2010) and consider the data generating process associated with the spatial lag model, shown in Eq. (3.35) to which we—assuming that ρ in absolute value is less than 1 and W is row-stochastic—have applied the well known infinite series expansion in Eq. (3.36) to express the inverse of $(I - \rho W)$

$$y = (I - \rho W)^{-1}X\beta + (I - \rho W)^{-1}\varepsilon \qquad (3.35)$$

$$(I - \rho W)^{-1} = I + \rho W + \rho^2 W^2 + \rho^3 W^3 + \dots \qquad (3.36)$$

$$y = X\beta + \rho WX\beta + \rho^2 W^2 X\beta + \rho^3 W^3 X\beta + \dots$$
$$+ \varepsilon + \rho W\varepsilon + \rho^2 W^2 \varepsilon + \rho^3 W^3 \varepsilon + \dots \qquad (3.37)$$

The model statement in Eq. (3.37) can be interpreted as indicating that the expected value of each observation y_i will depend on the mean plus a linear combination of values taken by neighbouring observations (areal units), scaled by the dependence parameters $\rho, \rho^2, \rho^3, \dots$

Consider the powers of the row-stochastic spatial weights matrix W (that is, $W^2, W^3, \ldots$) which appear in Eq. (3.37) where we assume that the rows of W are constructed to represent *first order* contiguous neighbours. Then the matrix W^2 will reflect *second order* contiguous neighbours, those that are neighbours to the first order neighbours. Since the neighbour of the neighbour (second order neighbour) to an observation i includes observation i itself, W^2 has positive elements on the main diagonal, when each observation has at least one neighbour. That is, higher order spatial lags can lead to a connectivity relation for an observation i such that $W^2 X\beta$ and $W^2\varepsilon$ will extract observations from the vectors $X\beta$ and ε that point back to the observation i itself. This is in contrast to the conventional independence relation in ordinary least squares regressions where the Gauss-Markov assumptions rule out dependence of ε_i on other observations $(j \neq i)$, by assuming zero covariance between observations i and j in the data generating process (LeSage and Pace 2010).

In standard least squares regression models of type (3.2) where the dependent variable vector contains *independent* observations, changes in observation i on the qth (non-constant) explanatory variable, which we denote by $\bar{X}_{iq}$, only influence observation y_i, so that the parameters have a straightforward interpretation as partial derivatives of the dependent variable with respect to the explanatory variable

$$\frac{\partial y_i}{\partial X_{jq}} = \begin{cases} \beta_q & \text{for } i = j \text{ and } q = 1, \ldots, Q - 1 \\ 0 & \text{for } j \neq i \text{ and } q = 1, \ldots, Q - 1. \end{cases} \tag{3.38}$$

The SAR model (and the spatial Durbin model) allows this type of change to influence y_i as well as other observations y_j with $j \neq i$. This type of impact arises due to the interdependence or connectivity between observations in the model. To see how this works, consider the spatial lag model expressed as shown in Eq. (3.39)

$$y = \sum_{q=1}^{Q} S_q(W)\bar{X}_q + V(W)\iota_n\beta_0 + V(W)\varepsilon \tag{3.39}$$

$$S_q(W) = V(W)(I\beta_q) \tag{3.40}$$

$$V(W) = (I - \rho W)^{-1} \tag{3.41}$$

where β_0 is the constant term parameter on ι_n, the n-by-1 vector of ones. Note that in the case of the SDM model $S_q(W) = V(W) \, (I\beta_q + W\gamma_q)$. For more details see LeSage and Pace (2009, 34 pp.).

To illustrate the role of $S_q(W)$, we rewrite the expansion of the data generating process in Eq. (3.39) as shown in Eq. (3.42)

$$
\begin{pmatrix} y_1 \\ y_2 \\ \vdots \\ y_n \end{pmatrix} = \sum_{q=1}^{Q-1} \begin{pmatrix} S_q(W)_{11} & S_q(W)_{12} & \cdots & S_q(W)_{1n} \\ S_q(W)_{21} & S_q(W)_{22} & \cdots & S_q(W)_{2n} \\ \vdots & \vdots & \vdots & \vdots \\ S_q(W)_{n1} & S_q(W)_{n2} & \cdots & S_q(W)_{nn} \end{pmatrix} \begin{pmatrix} \bar{X}_{1q} \\ \bar{X}_{2q} \\ \vdots \\ \bar{X}_{nq} \end{pmatrix} + V(W)\iota_n\beta_0 + V(W)\varepsilon.
$$

$$(3.42)$$

To make the role of $S_q(W)$ clear, consider the determination of a single dependent variable observation y_i

$$
y_i = \sum_{q=1}^{Q} [S_q(W)_{i1}\bar{X}_{1q} + S_q(W)_{i2}\bar{X}_{2q} + \cdots + S_q(W)_{in}\bar{X}_{nq}]
$$
$$
+ V(W)_i \iota_n \beta_0 + V(W)_i \varepsilon \qquad (3.43)
$$

where $S_q(W)_{ij}$ denotes the (i, j)th element of the matrix $S_q(W)$, and $V(W)_i$ the ith row of $V(W)$. It follows from Eq. (3.43) that—unlike to the case of the independent regression model—the derivative of y_i with respect to $\bar{X}_{jq}$ $(j \neq i)$ is potentially non-zero, taking a value determined by the (i, j)th element of the matrix $S_q(W)$, see LeSage and Pace (2010):

$$
\frac{\partial y_i}{\partial X_{jq}} = S_q(W)_{ij}. \qquad (3.44)
$$

In contrast to the least squares case, the derivative of y_i with respect to $\bar{X}_{iq}$ usually does not equal β_q, but results in an expression $S_q(W)_{ii}$ that measures the impact on the dependent observation i from a change in $\bar{X}_{iq}$

$$
\frac{\partial y_i}{\partial X_{iq}} = S_q(W)_{ii}. \qquad (3.45)
$$

Hence, a change to an explanatory variable in a single area (observation) can affect the dependent variable in other areas (observations). This is a logical consequence of the simultaneous spatial dependence structure in the spatial lag model. A change in the characteristics of neighbouring areal units can set in motion changes in the dependent variable that will impact the dependent variable in neighbouring areas. These impacts will diffuse through the system of areas.

Since the partial derivatives take the form of an n-by-n matrix and since there are Q–1 non-constant explanatory variables, this results in $(Q - 1)n^2$ partial derivatives which provides an overwhelming amount of information. LeSage and Pace (2009, pp. 36–37) suggest summarising these partial derivatives. In particular, they propose averaging all the column or row sums of $S_q(W)$ to arrive at the *average total impact* or *effect*, averaging the main diagonal elements of this matrix to arrive at the *average direct impact* or *effect*, and averaging the off-diagonal elements of $S_q(W)$ to arrive at the *average indirect impact* or *effect*. This latter summary measure reflects what are commonly thought of as *spatial spillovers*, or impacts falling on areas other than the own-area.

One applied illustration that uses these scalar summary impact estimates can be found in Fischer et al. (2009b). The application considers the direct, indirect and total impacts of changes in human capital on labour productivity levels in European regions. A number of other applications can be found in LeSage and Pace (2009) in a wide variety of application contexts.

For inference regarding the significance of these impacts, one needs to determine their empirical or theoretical distributions. Since the impacts reflect a nonlinear combination of the parameters ρ and $\bar{\beta}$ in the case of the SAR model, working with the theoretical distribution is not very convenient. Given the model estimates as well as the associated variance–covariance matrix along with the knowledge that the ML estimates are (asymptotically) normally distributed, one can simulate the parameters ρ and $\bar{\beta}$ (and γ in the case of the SDM model). These empirically simulated magnitudes can be used in the expressions for the scalar summary measures to generate an empirical distribution of the scalar impact measures (LeSage and Pace 2009, 2010).

An illustration of a simulation approach to determining measures of dispersion for these scalar summary impact estimates can be found in Fischer et al. (2009b). Another illustration is given in LeSage and Fischer (2008) in the context of Bayesian model averaging methods.

Part II
The Analysis of Spatial Interaction Data

In Part II of this book we direct attention to the analysis of spatial interaction data, that is, observations, say with $i, j = 1, ..., n$, on random variables $Y(i, j)$ each of which corresponds to movements of people (cars, commodities, telephone calls etc.) between spatial locations i and j, where each location is both origin and destination of interaction. The locations may be point locations, or alternatively areas (zones), or a mixture of these, for example, if we are interested in the use of shopping facilities such as shopping centres, the origins are likely to be residential areas, while the destinations will be the shopping centres which are fixed point locations.

The focus is on spatial interaction models of the gravity type. These models typically rely on three types of factors to explain mean interaction frequencies between origins and destinations of interaction: (i) origin-specific factors that characterise the ability of origins to generate flows, (ii) destination-specific factors that represent the attractiveness of destinations, and (iii) origin-destination factors that characterise the way spatial separation of origins from destinations constrains or impedes the interaction (Fischer 2010).

Such models are relevant in fields as international and interregional trade, transportation research, population migration research, shopping and commuting behaviour, journey-to-work and communication studies.

Chapter 4 introduces the general spatial interaction model that asserts a multiplicative relationship between mean interaction frequencies and the effects of origin, destination, and spatial separation. Particular emphasis is laid on Poisson spatial interaction model specifications that have served as the workhorse in spatial interaction analysis. The final chapter directs attention to the issue of spatial dependence in origin-destination flows, and discusses approaches to dealing with this problem.

Keywords Spatial interaction data · General spatial interaction model of the gravity type · Power and exponential deterrence functions · Poisson spatial interaction model specifications · Spatial dependence in the origin-destination flows · Econometric extensions to the independence (log-normal) model · Spatial filtering methodology · Maximum likelihood estimation

Chapter 4
Models and Methods for Spatial Interaction Data

Abstract The phenomenon of interest in this chapter may be described in most general terms as interactions between populations of actors and opportunities distributed over some relevant geographic space. Such interactions may involve movements of individuals from one location to another, such as daily traffic flows in which case the relevant actors are individuals such as commuters (shoppers) and the relevant opportunities are their destinations such as jobs (or stores). Similarly, one may consider migration flows, in which case the relevant actors are migrants (individuals, family units, firms) and the relevant opportunities are their possible new locations. Interactions may also involve flows of information such as telephone calls or electronic messages. Here the callers or message senders may be relevant actors, and the possible receivers of calls or electronic messages may be considered as the relevant opportunities (Sen and Smith 1995, Gravity models of spatial interaction behavior. Springer, Berlin pp. 18–19).

Keywords Origin–destination flow data · General spatial interaction model of the gravity type · Origin-specific variables · Destination-specific variables · Separation functions · Poisson spatial interaction model specifications · Overdispersion · The negative binomial model of spatial interaction

4.1 Visualising and Exploring Spatial Interaction Data

Suppose that we have a spatial system consisting of n' origins and n destinations, and that $y(i,j)$ is the number of observed origin–destination (OD) flows from origin location i ($i = 1, \ldots, n'$) to destination location j ($j = 1, \ldots, n$). These flows are positive integers including zero that may be displayed in the form of an interaction matrix Y as in (4.1):

M. M. Fischer and J. Wang, *Spatial Data Analysis*,
SpringerBriefs in Regional Science, DOI: 10.1007/978-3-642-21720-3_4,
© Manfred M. Fischer 2011

$$Y = \begin{bmatrix} y(1,1) & \cdots & y(1,j) & \cdots & y(1,n) \\ \vdots & & \vdots & & \vdots \\ y(i,1) & \cdots & y(i,j) & \cdots & y(i,n) \\ \vdots & & \vdots & & \vdots \\ y(n',1) & \cdots & y(n',j) & \cdots & y(n',n) \end{bmatrix}. \tag{4.1}$$

In most applications, for example, migration modelling, $n = n'$ and hence Y is a square matrix with n^2 elements. Even though in some applications—for example, shopping trips from residential areas i to individual shopping centres j—the number of origin and destination locations may differ, and thus Y will be not square, for notational simplicity we restrict the discussion to the case of a square matrix where each origin is also a destination (that is, $n' = n$). The elements on the main diagonal, $y(i,i)$ for $i = 1, \ldots, n$, represent intrazonal flows. But note that there are situations where intrazonal flows are difficult to record, and there are also situations—for example, flows by airlines between airports—where intrazonal flows have no meaning. The ith row of the matrix describes outflows from location i to each of the n destination locations, while the jth column of the matrix describes inflows from each of the n origin locations into destination location j.

The basic mode of representing OD flows $\{y(i,j): i,j = 1, \ldots, n\}$ centres around maps. One way of displaying the flows on a map is to draw line segments between each pair of centroids representing spatial units. To show the volumes of flows, the segments may be coloured, for example, or drawn with varying thickness. Since flows are directional (or in other words $y(i,j) \neq y(j,i)$), it is necessary to show them in both directions using arrows.

The problem of representing spatial interaction data lies in the number of locations. The map representation fails if the number of locations is large because the display becomes too cluttered. For a spatial interaction system with 100 locations, for example, there are 9,900 directional node pairs. There are two possible solutions to resolve the map clutter problem: first, to apply interactive parameter manipulations such as thresholding and filtering to reduce the visual complexity, and second, to use a visual matrix representation where the links are represented by squares tiled on the display, but the easy interpretation and context provided by the map is lost (Fischer 2000, p. 41).

Some exploratory tools attempt to uncover evidence of hierarchical structure in spatial interaction data. If we are dealing with a flow matrix in which the origin and destination locations are urban areas, some of these will be located at higher positions in the urban hierarchy than others. Some urban areas will thus come to dominate others, not in a simple geographical sense but in terms of functional linkages. Bailey and Gatrell (1995, p. 347), suggest to use the flow data in a spatial interaction matrix as given by (4.1) to reveal patterns of dominance. A simple way to do this graphically is to represent each location as a dot on a topological map and to link i to j with a directed arc if $y(i,j) > y(i,j')$ for $j' = 1, \ldots, n$ and $j \neq j'$ and if location j is larger than location i where size may be measured in terms of the column totals in matrix (4.1). This procedure can be made more sophisticated, but it provides some evidence of hierarchical structure in the form of a directed graph (Bailey and Gatrell 1995, p. 347).

4.2 The General Spatial Interaction Model

Let us move on now to consider more formal models for spatial interaction data. We consider, mathematically, a situation in which a series of observations $\{y(i,j): i,j = 1,\ldots,n\}$ on random variables $Y(i,j)$ is given, each of which corresponds to movements of people (cars, commodities or telephone calls) between origin and destination locations i and j. The $Y(i,j)$ are assumed to be independent random variables. They are sampled from a specified probability distribution that is dependent upon some mean, say $\mu(i,j)$.

We are interested in modelling observed origin–destination flows according to a statistical model of the general form

$$Y(i,j) = \mu(i,j) + \varepsilon(i,j) \quad i,j = 1,\ldots,n \qquad (4.2)$$

where $\mu(i,j) = E[Y(i,j)]$ is the expected mean interaction frequency from i to j, and $\varepsilon(i,j)$ is an error about the mean.

One of the most common classes of models for $\mu(i,j)$—known as spatial interaction or gravity models—relies on three types of factors: (i) origin-specific factors that characterise the ability of the origin locations to produce or generate flows, (ii) destination-specific factors that represent the attractiveness of destinations, and (iii) origin–destination factors that characterise the way spatial separation of origins from destinations constrains or impedes the interaction (see, for example, Fischer 2010). Spatial interaction models essentially assert a multiplicative relationship between mean interaction frequencies and the effects of origin, destination, and separation factors, respectively. The mean interaction frequencies between origin i and destination j are modelled by

$$\mu(i,j) = C\,A(i)\,B(j)\,S(i,j) \quad i,j = 1,\ldots,n \qquad (4.3)$$

where $A(i)$ and $B(j)$ are called origin-specific and destination-specific factors (functions), respectively. $S(i,j)$ is a function of some measure of separation between locations i and j, and C denotes some constant of proportionality (see Fischer and Griffith 2008).

Alternative forms of this *general spatial interaction model* can be specified by imposing different constraints on $\mu(i,j)$. For the relationship between the constraints and the constant, see, for example, Ledent (1985). In the *unconstrained* case the only condition specified is that the estimated mean interaction frequencies equal the observed mean interaction frequencies

$$\sum_{i=1}^{n}\sum_{j=1}^{n}\mu(i,j) = \sum_{i=1}^{n}\sum_{j=1}^{n}y(i,j). \qquad (4.4)$$

In the *origin-constrained* and the *destination-constrained* cases the estimated outflows from each location (origin constrained) and the estimated inflows to each location (destination constraint) have to match the observed outflow totals $y(i,\cdot)$ and the observed inflow totals $y(\cdot,j)$ respectively, that is

$$y(i,\cdot) = \sum_{j=1}^{n}y(i,j) = \sum_{j=1}^{n}\mu(i,j) \qquad (4.5)$$

$$y(\cdot,j) = \sum_{i=1}^{n}y(i,j) = \sum_{i=1}^{n}\mu(i,j). \qquad (4.6)$$

Finally, in the *doubly constrained* case both the estimated inflows and outflows must equal their observed counterparts, that is both Eqs. (4.5) and (4.6) must be satisfied. In the sections that follow we confine our attention to the unconstrained case where we have no a priori information on the marginal totals of the spatial interaction matrix if future flows have to be predicted.

4.3 Functional Specifications and the Method of Ordinary Least Squares Regression

Equation (4.3) is a very general version of a spatial interaction model. This distinguishes the model from others in which the $A(i), B(j)$ and $S(i,j)$ terms are more specific. The exact functional form of these terms is subject to varying degrees of conjectures. But there is wide agreement that the origin and destination factors are best given by power functions (see Fotheringham and O'Kelly 1989, p. 10):

$$A(i) = A(i,\beta) = (A_i)^{\beta} \quad i = 1,\ldots,n \qquad (4.7)$$

$$B(j) = B(j,\gamma) = (B_j)^{\gamma} \quad j = 1,\ldots,n \qquad (4.8)$$

where $A(i)$ represents some appropriate variable measuring the propulsiveness of origin i, and $B(j)$ some appropriate variable measuring the attractiveness of destination j in a specific spatial interaction context. The product $A(i)B(j)$ can be interpreted simply as the number of distinct (i,j)-interactions that are possible. Thus, for origin–destination pairs (i,j) with the same level of separation, it follows from Eq. (4.3) that mean interaction levels are proportional to the number of possible interactions between such (i,j)-pairs. The exponents, β and γ, indicate the origin and destination effects, respectively, and are treated as statistical parameters to be estimated.

If more than one origin and one destination variable are relevant in a specific context the above specifications may be extended to

$$A(i) = A(i, \beta) = \prod_{q=1}^{Q} (A_{iq})^{\beta_q} \quad i = 1, \ldots, n \tag{4.9}$$

$$B(j) = B(j, \gamma) = \prod_{r=1}^{R} (B_{jr})^{\gamma_r} \quad j = 1, \ldots, n \tag{4.10}$$

where $A_{iq}(q = 1, \ldots, Q)$ and $B_{jr}(r = 1, \ldots, R)$ represent sets of relevant (positive) origin-specific and destination-specific variables, respectively. The exponents $\beta = (\beta_q: q = 1, \ldots, Q)$ and $\gamma = (\gamma_r: r = 1, \ldots, R)$ are parameters to be estimated.

The separation function $S(i, j)$ constitutes the very core of spatial interaction models. Hence, a number of alternative specifications have been proposed in the literature (for a discussion see Sen and Smith 1995, pp. 92–99). One prominent example is the power function specification given by

$$S(i, j) = S[D(i, j), \theta] = [D(i, j)]^{-\theta} \quad i, j = 1, \ldots, n \tag{4.11}$$

for any positive scalar separation measure $D(i, j)$, and positive sensitivity parameter θ that has to be estimated. Note that this type of functional specification is questionable for small separation values. Indeed, since $D \to 0$ implies that $D^{-\theta} \to \infty$ for all $\theta > 0$, it follows that mean interaction frequencies between origin and destination locations with very small values of interaction separation must be overwhelmingly larger than other mean interaction levels (Sen and Smith 1995, p. 94). But since this type of interaction behaviour is generally not observed, it has been widely recognised that power separation functions are generally not the first choice for modelling spatial interactions involving small separation values (see Fotheringham and O'Kelly 1989, pp. 12–13).

The other specification that has generated a great deal of interest in the literature is the exponential separation function

$$S(i, j) = S[D(i, j), \theta] = \exp[-\theta D(i, j)] \quad i, j = 1, \ldots, n \tag{4.12}$$

in which θ may again be interpreted as a positive separation sensitivity parameter. As $D(i, j)$ increases, $\exp[-\theta D(i, j)]$ decreases. It decreases faster the larger θ. But it is important to emphasise that—unlike the power specification (4.11)—the parameter θ must be a *dimensional parameter* with specific value depending on the choice of units for interaction distance (Sen and Smith 1995, p. 95). In other words, Eq. (4.12) is a meaningful parameterised function with respect to distance if and only if all transformations of measurement units are taken to be absorbed in the definition of θ. Note that the interest in exponential separation functions centres on their theoretical significance from a behavioural point of view (Sen and Smith 1995).

The separation (also termed deterrence) function $S(i, j)$ reflects the way in which spatial separation constrains or impedes movement across space. In general we will refer to this as distance between an origin i and a destination j, and denote it as $D(i,j)$. At relatively large scales of geographical inquiry this might be simply the great circle distance separating an origin from a destination area measured in terms of the distance between their respective centroids. In other cases, it might be transportation or travel time, cost of transportation, perceived travel time or any other sensible measure such as political distance, language distance or cultural distance measured in terms of nominal or categorical attributes.

To allow for the possibility of multiple measures of spatial separation, the power function specification in Eq. (4.11) can be extended to the following class of multivariate power deterrence function

$$S(i,j) = \prod_{k=1}^{K} \left[D^{(k)}(i,j) \right]^{-\theta_k} \quad i,j = 1,\ldots,n \tag{4.13}$$

for any set of relevant separation measures $\{D^{(k)}(i,j): k = 1,\ldots,K\}$, with the corresponding separation sensitivity vector $\theta = (\theta_k: k = 1,\ldots,K)$.

Similarly, for any set of relevant separation measures $D^{(k)}$, one may extend the exponential separation function given by Eq. (4.12) to the multivariate case

$$S(i,j) = \exp\left\{ -\sum_{k=1}^{K} \theta_k D^{(k)}(i,j) \right\} \quad i,j = 1,\ldots,n \tag{4.14}$$

with the separation sensitivity vector, $\theta = (\theta_k: k = 1,\ldots,K)$.

Incorporating Eq. (4.13) with specifications (4.9) and (4.10) into Eq. (4.3) yields the (*multivariate*) *power spatial interaction model*

$$\mu(i,j) = C \prod_{q=1}^{Q} (A_{iq})^{\beta_q} \prod_{r=1}^{R} (B_{jr})^{\gamma_r} \prod_{k=1}^{K} [D^{(k)}(i,j)]^{-\theta_k} \quad i,j = 1,\ldots,n. \tag{4.15}$$

And incorporating Eq. (4.14) with specifications (4.9) and (4.10) into Eq. (4.3) leads to the (*multivariate*) *exponential spatial interaction model*

$$\mu(i,j) = C \prod_{q=1}^{Q} (A_{iq})^{\beta_q} \prod_{r=1}^{R} (B_{jr})^{\gamma_r} \exp\left\{ -\sum_{k=1}^{K} \theta_k D^{(k)}(i,j) \right\} \quad i,j = 1,\ldots,n. \tag{4.16}$$

Having chosen a particular functional form for $\mu(i,j)$ such as, for example, the multivariate exponential spatial interaction model (4.16), one can go to the problem of modelling a set of observed flows $y(i,j)$. Recall that we think of these as observations on random variables $Y(i,j)$ with mean value $\mu(i,j)$. Incorporating (4.16) for $\mu(i,j)$ into Eq. (4.2) yields

$$Y(i,j) = C \prod_{q=1}^{Q} (A_{iq})^{\beta_q} \prod_{r=1}^{R} (B_{jr})^{\gamma_r} \exp\left\{ -\sum_{k=1}^{K} \theta_k D^{(k)}(i,j) \right\} + \varepsilon(i,j) \tag{4.17}$$

for $i, j = 1, \ldots, n$. From a statistical point of view, fitting this kind of spatial interaction model to observed data is a question of estimating the unknown parameters $\beta = (\beta_1, \ldots, \beta_Q)'$, $\gamma = (\gamma_1, \ldots, \gamma_R)'$ and $\theta = (\theta_1, \ldots, \theta_K)'$ and the scalar constant.

The Method of Ordinary Least Squares Regression As a first approach it is tempting to take logarithms of this model for $\mu(i,j)$, and write it in a linear form (Bailey and Gatrell 1995, p. 352) as

$$\ln Y(i,j) = \ln C + \sum_{q=1}^{Q} \beta_q A_{iq} + \sum_{r=1}^{R} \gamma_r B_{jr} - \sum_{k=1}^{K} \theta_k D^{(k)}(i,j) + \varepsilon'(i,j) \quad (4.18)$$

for $i, j = 1, \ldots, n$ with

$$\varepsilon'(i,j) \sim \mathcal{N}(0, \sigma^2) \quad (4.19)$$

and then proceed to estimate the parameters by ordinary least squares regression of the observations $y(i,j)$ on $A(i), B(j)$ and $D^{(k)}(i,j)$.

But such an approach suffers from two major drawbacks. *First*, the regression produces estimates of the parameters based on logarithms of the interaction and the explanatory variables, not of the interactions and variables themselves. One of the effects of this is to underpredict large origin–destination flows, and to under-predict the total number of flows (see Flowerdew and Aitkin 1982).

Second, estimating the parameters by the log-additive regression model given by Eqs. (4.18) and (4.19) would only be justified statistically if we believed that flows $Y(i,j)$ were independent and log-normally distributed about their mean value with a constant variance. Such an assumption, however, is patently not valid since origin–destination flows are discrete counts whose variance is very likely to be proportional to their mean value. Least squares assumptions ignore the true integer nature of the origin–destination flows and approximate a discrete valued process by an almost certainly misrepresentative continuous distribution (see Bailey and Gatrell 1995, p. 353). Consequently, ordinary least squares regression estimates and their standard errors can be seriously distorted.

4.4 The General Poisson Spatial Interaction Model

Maximum likelihood estimation of the parameters under more realistic distributional assumptions is generally considered as more appropriate approach. The most common assumption is that the $Y(i, j)$ follow independent Poisson distributions with expected values $\mu(i,j) = A(i)B(j)S(i,j)$. This assumption is also open to question, since origin–destination flows are not strictly independent and, more-over, a Poisson distribution may not adequately reflect the degree of variation present in many real world datasets, since in many application contexts individuals may tend to "flow" in groups rather than as individuals. Nevertheless, this assumption is generally considered as providing reasonable parameter estimates, at least in the first instance (Bailey and Gatrell 1995, p. 353). The alternative would be to use some distributional assumption more able to reflect overdispersion, an issue that we will discuss later in Sect. 4.6.

The primary equation of the Poisson model of spatial interaction is the Poisson density, or more formally the Poisson probability mass function

$$\text{Prob}[Y(i,j) = y(i,j)|\mu(i,j)]$$
$$= \frac{\exp[-\mu(i,j)][\mu(i,j)]^{y(i,j)}}{y(i,j)!} \quad y(i,j) = 0,1,2,\ldots \text{ and } i,j = 1,\ldots,n \quad (4.20)$$

where μ is the intensity or rate parameter that is usually parameterised as

$$\mu(i,j) = E[y(i,j)|A(i),B(j),S(i,j)]$$
$$= \exp[A(i,\beta)B(j,\gamma)S[D(i,j),\theta]] \quad i,j = 1,\ldots,n. \quad (4.21)$$

This specification (4.21) is called the *exponential mean* parameterisation that has the advantage of ensuring that $\mu > 0$. The model consisting of Eqs. (4.20) and (4.21) is referred to as *general Poisson spatial interaction model* since the origin, destination and separation factors are not yet specified (see Sect. 4.3 for the functional specification of these terms). Note that $\mu(i,j)$ is a deterministic function of $A(i)$, $B(j)$ and $S(i,j)$, and the randomness in the model comes from the Poisson specification of $y(i,j)$.

This model shows the well-known equidispersion (equality of mean and variance) property of the Poisson distribution, that is

$$var[y(i,j)|A(i),B(j),S(i,j)]$$
$$= E[y(i,j)|A(i),B(j),S(i,j)] = \mu(i,j) \quad i,j = 1,\ldots,n. \quad (4.22)$$

It also implies the conditional mean to have a multiplicative form given by $E[y(i,j)|A(i),B(j),S(i,j)] = A(i,\beta)B(j,\gamma)S[D(i,j),\theta]$. The density of the Poisson spatial interaction model for a single observation (i,j) is

$$f[y(i,j)|A(i),B(j),S(i,j),\beta,\gamma,\theta]$$
$$= \frac{\exp\{-\exp[A(i,\beta)B(j,\gamma)S[D(i,j),\theta)]]\}\exp[A(i,\beta)B(j,\gamma)S[D(i,j),\theta)]]}{y(i,j)!}. \quad (4.23)$$

The Poisson specification of the spatial interaction model shows some interesting features. *First*, it is analogous to the familiar regression specification in many ways. In particular, $E[y(i,j)|A(i),B(j),S(i,j)] = \mu(i,j)$ for $i,j = 1,\ldots,n$. Moreover, parameter estimation is straightforward and may be done by maximum likelihood (see Sect. 4.5). *Second*, the 'zero problem', $y(i,j) = 0$, is a natural outcome of the Poisson specification. In contrast to the logarithmic regression specification there is no need to truncate an arbitrary continuous distribution.

The integer property of the outcomes $y(i,j)$ is handled directly (Fischer et al. 2006). Finally, it is worth noting that the Poisson spatial interaction model is simply a non-linear regression.

4.5 Maximum Likelihood Estimation of the Poisson Spatial Interaction Model

Since each $Y(i,j)$ has a Poisson distribution and all the $Y(i,j)$ are independent, the likelihood function is given as

$$L(\beta,\gamma,\theta) = \prod_{i=1}^{n}\prod_{j=1}^{n} \exp\left\{-A(i,\beta)B(j,\gamma)S[D(i,j),\theta]\left[A(i,\beta)B(j,\gamma)\right.\right.$$

$$\left.\left. S[D(i,j),\theta]\right]^{y(i,j)} \frac{1}{y(i,j)!}\right\} \tag{4.24}$$

and the values $\hat{A}(i,\beta), \hat{B}(j,\gamma)$ and $\hat{\theta}$ of $A(i,\beta), B(j,\gamma)$ and θ that maximise (4.24) are called maximum likelihood estimates. If the likelihood function is maximised by $\{\hat{A}(i,\beta), \hat{B}(j,\gamma), \hat{S}[D(i,j),\theta]\}$, so will its logarithm, the log-likelihood function, which is seen from Eq. (4.24) (see Sen and Smith 1995, pp. 359–361) to be for the case of the exponential deterrence function (4.14)

$$\ln L(\beta,\gamma,\theta) = \sum_{i=1}^{n}\sum_{j=1}^{n}\{-A(i,\beta)B(j,\gamma)\exp[-\theta D(i,j)]$$

$$+y(i,j)[\ln A(i,\beta) + \ln B(j,\gamma) - \theta D(i,j)] - \ln[y(i,j)!]\}$$

$$= \sum_{i=1}^{n}\sum_{j=1}^{n}\{-A(i,\beta)B(j,\gamma)\exp[-\theta D(i,j)]\}$$

$$+ \sum_{i=1}^{n}y(i,\cdot)\ln A(i,\beta) + \sum_{j=1}^{n}y(\cdot,j)\ln B(j,\gamma)$$

$$-\theta[D(i,j)y(i,j)] - \sum_{i=1}^{n}\sum_{j=1}^{n}\ln[y(i,j)!] \tag{4.25}$$

with

$$y(i,\cdot) = \sum_{j=1}^{n}y(i,j) \quad i=1,\ldots,n \tag{4.26}$$

$$y(\cdot,j) = \sum_{i=1}^{n}y(i,j) \quad j=1,\ldots,n. \tag{4.27}$$

The partial derivatives of $\ln L(\beta,\gamma,\theta)$ are given by

$$\frac{\partial \ln L(\beta, \gamma, \theta)}{\partial \beta} = \sum_{j=1}^{n} \left\{ -B(j, \gamma) \exp[-\theta D(i,j)] + \frac{y(i,j)}{A(i, \beta)} \right\} \frac{\partial A(i, \beta)}{\partial \beta}$$

$$= \frac{[y(i, \cdot) - \mu(i, \cdot)]}{A(i, \beta)} \frac{\partial A(i, \beta)}{\partial \beta} \quad \text{for} \ \ i = 1, \dots, n \tag{4.28}$$

$$\frac{\partial \ln(\beta, \gamma, \theta)}{\partial \gamma} = \sum_{i=1}^{n} \left\{ -A(i, \beta) \exp[-\theta D(i,j)] + \frac{y(i,j)}{B(j, \gamma)} \right\} \frac{\partial B(j, \gamma)}{\partial \gamma}$$

$$= \frac{[y(\cdot, j) - \mu(\cdot, j)]}{B(j, \gamma)} \frac{\partial B(j, \gamma)}{\partial \gamma} \quad \text{for} \ \ j = 1, \dots, n \tag{4.29}$$

$$\frac{\partial \ln L(\beta, \gamma, \theta)}{\partial \theta} = \sum_{i=1}^{n} \sum_{j=1}^{n} \{ -D(i,j)A(i, \beta)B(j, \gamma) \exp[-\theta D(i,j)]$$

$$+ y(i,j)D(i,j) \} = \sum_{i=1}^{n} \sum_{j=1}^{n} D(i,j)[y(i,j) - \mu(i,j)] \tag{4.30}$$

with

$$\mu(i, \cdot) = \sum_{j=1}^{n} \mu(i,j) \quad i = 1, \dots, n \tag{4.31}$$

$$\mu(\cdot, j) = \sum_{i=1}^{n} \mu(i,j) \quad j = 1, \dots, n. \tag{4.32}$$

Maximum likelihood estimates may be found by maximising $\ln L(\phi = (\beta, \gamma, \theta))$ directly using iterative procedures, usually gradient algorithms such as Newton–Raphson. Alternatively, one could set the partial derivatives of $\ln L(\phi)$ (i.e. Eqs. (4.28)–(4.30)) equal to zero and solve the resultant equations

$$\mu(i, \cdot) = y(i, \cdot) \quad i = 1, \dots, n \tag{4.33}$$

$$\mu(\cdot, j) = y(\cdot, j) \quad j = 1, \dots, n \tag{4.34}$$

$$\sum_{i=1}^{n} \sum_{j=1}^{n} D(i,j)\mu(i,j) = \sum_{i=1}^{n} \sum_{j=1}^{n} D(i,j)y(i,j). \tag{4.35}$$

Convergence is guaranteed because the log-likelihood function is globally concave (Sen and Smith 1995, pp. 359–361).

Finally, it is worth noting that maximum likelihood estimation is not restricted to the exponential separation function specification of the general Poisson spatial interaction model. Any continuously differentiable separation function $S[D(i,j), \theta]$ will do, although the partial derivatives with respect to θ,

$$\frac{\partial \ln L(\beta, \gamma, \theta)}{\partial \theta} = \frac{\partial S[D(i,j), \theta]}{\partial \theta} \frac{1}{S[D(i,j), \theta]}$$

$$\sum_{i=1}^{n} \sum_{j=1}^{n} \{-A(i, \beta)B(j, \gamma)S[D(i,j), \theta] + y(i,j)\} \qquad (4.36)$$

would not necessarily produce the simple equations akin to Eqs. (4.33)–(4.35) (Sen and Smith 1995, p. 361).

4.6 A Generalisation of the Poisson Model of Spatial Interaction

The general Poisson model of spatial interaction is the workhorse in spatial interaction data analysis. One deficiency of this model specification, however, is that for origin–destination flow data the variance usually exceeds the mean, a feature called *overdispersion*. Overdispersion has qualitatively similar consequences to the failure of homoscedasticity in the linear regression model. Provided that the conditional mean is correctly specified, that is, Eq. (4.21) holds, Poisson maximum likelihood estimation is still consistent. It is nonetheless important to control for overdispersion. Even in the simplest spatial interaction settings large overdispersion leads to grossly deflated standard errors and grossly deflated t-statistics in the usual ML output (Cameron and Trivedi 2005, p. 670).

The standard parametric model to account for overdispersion is the negative binomial. Suppose that the distribution of the random variables $Y(i,j)$ is Poisson conditional on the parameter λ so that the probability mass function

$$f[y(i,j)|\lambda(i,j)] = \frac{\exp[-\lambda(i,j)]\lambda(i,j)^{y(i,j)}}{y(i,j)!} \qquad (4.37)$$

for $y(i,j) = 0, 1, 2, \ldots$ and $i, j = 1, \ldots, n$. Suppose now that the parameter λ is random, rather than being a completely deterministic function of $A(i, \beta)$, $B(j, \gamma)$ and $S[D(i,j), \theta]$. In particular, let $\lambda = \mu \upsilon$, where μ is a deterministic function of $A(i, \beta)$, $B(j, \gamma)$ and $S[D(i,j), \theta]$, for example as given by Eq. (4.21), and $\upsilon > 0$ is *iid* with density $g(\upsilon|\alpha)$. This is an example of unobserved heterogeneity, as different observations may have different λ (heterogeneity) but part of this difference is due to a random (unobserved) component υ. Note that $E[\lambda|\mu] = \mu$ if $E[\upsilon] = 1$ so that the interpretation of the parameters β, γ and θ stays as in the Poisson spatial interaction model.

The marginal density of $y(i,j)$, unconditional on the random parameter υ but conditional on the deterministic parameters μ and α, is obtained by integrating out υ. This gives

$$h[y(i,j)|\mu(i,j),\alpha] = \int_0^\infty f[(i,j),\upsilon]g(\upsilon|\alpha)\,\mathrm{d}\upsilon \qquad (4.38)$$

with $y(i,j) = 0, 1, 2, \ldots$ and $i,j = 1, \ldots, n$, where $g(\upsilon|\alpha)$ is called the mixing distribution and α denotes the unknown parameter of the mixing distribution. The integration defines an 'average' distribution. If $f[y(i,j)|\mu(i,j)]$ is the Poisson density and

$$g(\upsilon) = \frac{\upsilon^{\delta-1}\exp(-\upsilon\delta)\delta^\delta}{\Gamma(\delta)} \qquad (4.39)$$

where $\upsilon, \delta > 0$ is the gamma density with $E[\upsilon] = 1$ and $\mathrm{var}[\upsilon] = \delta^{-1}$, we obtain the negative binomial as a mixture density

$$h[y(i,j)|\mu(i,j),\delta]$$

$$= \int_0^\infty \frac{\exp[-\mu(i,j)\upsilon][\mu(i,j)\upsilon]^{y(i,j)}}{y(i,j)!}\frac{\upsilon^{\delta-1}\exp(-\upsilon\delta)\delta^\delta}{\Gamma(\delta)}\mathrm{d}\upsilon$$

$$= \frac{\mu(i,j)^{y(i,j)}\delta^\delta\Gamma[y(i,j)+\delta]}{\Gamma(\delta)y(i,j)![\mu(i,j)+\delta]^{y(i,j)+\delta}}$$

$$= \frac{\Gamma[\alpha^{-1}+y(i,j)]}{\Gamma(\alpha^{-1})\Gamma[y(i,j)+1]}\left(\frac{\alpha^{-1}}{\alpha^{-1}+\mu(i,j)}\right)^{1/\alpha}\left(\frac{\mu(i,j)}{\mu(i,j)+\alpha^{-1}}\right)^{y(i,j)} \qquad (4.40)$$

for $y(i,j) = 0, 1, 2, \ldots$ and $i,j = 1, \ldots, n$, where $\alpha = \delta^{-1}$ and $\Gamma(\cdot)$ denotes the gamma integral that specialises to a factorial for an integer argument, and the third line follows after some algebra and use of the definition of the gamma function (Cameron and Trivedi 2005, pp. 673–676).

Equation (4.40) along with Eqs. (4.37)–(4.39) is referred to as *negative binomial model of spatial interaction*. The first two moments of this model are

$$E[y(i,j)|\mu(i,j),\alpha] = \mu(i,j) \qquad (4.41)$$

$$\mathrm{var}[y(i,j)|\mu(i,j),\alpha] = \mu(i,j)[1+\alpha\mu(i,j)]. \qquad (4.42)$$

Thus, the model allows for overdispersion (that is $\alpha > 0$), and underdispersion (that is $\alpha < 0$), with $\alpha = 0$ reducing to the Poisson model specification (4.20)–(4.21).

Estimation of the model may proceed with maximum likelihood. The log-likelihood function for exponential mean parameterisation $\mu(i,j)$ is given (Cameron and Trivedi 1998, p. 71) by

$$\ln L(\alpha, \beta, \gamma, \theta) = \sum_{i=1}^{n} \sum_{j=1}^{n} \left\{ \ln \left[\frac{\Gamma[y(i,j) + \alpha^{-1}]}{\Gamma(\alpha^{-1})} \right] - \ln y(i,j)! \right.$$

$$- \left[y(i,j) + \alpha^{-1} \right] \ln \left[1 + \alpha \exp[A(i,\beta)B(j,\gamma)S[D(i,j),\theta]] \right]$$

$$\left. + y(i,j) \ln \alpha + y(i,j) \left[A(i,\beta)B(j,\gamma)S[D(i,j),\theta] \right] \right\}.$$

$$(4.43)$$

An applied illustration is given in Fischer et al. (2006). The authors have found the negative binomial spatial interaction model to be very useful. It appears to have the flexibility for providing a good fit, and it does so in part because the quadratic variance specification is a good approximation in many application contexts. Finally, it is worth noting that several software packages offer the negative binomial model as a standard option. Alternatively, one can use a ML routine with user-provided log-likelihood function and possibly derivatives (Cameron and Trivedi 1998, p. 27).

Chapter 5
Spatial Interaction Models and Spatial Dependence

Abstract Spatial interaction models of the types discussed in the previous chapter take the view that inclusion of a spatial separation function between origin and destination locations is adequate to capture any spatial dependence in the sample data. LeSage and Pace (J Reg Sci 48(5):941–967, 2008), and Fischer and Griffith (J Reg Sci 48(5):969–989, 2008) provide theoretical as well as an empirical motivation that this may not be adequate to model potentially rich patterns that can arise from spatial dependence. In this chapter we consider three approaches to deal with spatial dependence in origin–destination flows. Two approaches incorporate spatial correlation structures into the independence (log-normal) spatial interaction model. The first specifies a (first order) spatial autoregressive process that governs the spatial interaction variable (see LeSage and Pace (J Reg Sci 48(5):941–967, 2008)). The second approach deals with spatial dependence by specifying a spatial process for the disturbance terms, structured to follow a (first order) spatial autoregressive process. In this framework, the spatial dependence resides in the disturbance process (see Fischer and Griffith (J Reg Sci 48(5):969–989, 2008). A final approach relies on using a spatial filtering methodology developed by Griffith (Spatial autocorrelation and spatial filtering, Springer, Berlin, Heidelberg and New York, 2003) for area data, and leads to eigenfunction based spatial filtering specifications of both the log-normal and the Poisson spatial interaction model versions (see Fischer and Griffith (J Reg Sci 48(5):969–989, 2008)).

Keywords Origin–destination flow data · Independence (log-normal) spatial interaction model · Origin-based spatial dependence · Destination-based spatial dependence · Origin-to-destination based spatial dependence · Econometric extensions to the independence spatial interaction model · Spatial filtering methodology · Eigenfunction based spatial filter model specifications

M. M. Fischer and J. Wang, *Spatial Data Analysis*,
SpringerBriefs in Regional Science, DOI: 10.1007/978-3-642-21720-3_5,
© Manfred M. Fischer 2011

5.1 The Independence (Log-Normal) Spatial Interaction Model in Matrix Notation

Recall that Y denotes an n-by-n matrix of origin–destination flow data as shown in (4.1) where the n rows represent different origins and the n columns different destinations. The elements on the main diagonal of the matrix represent flows within the locations, and we use $N = n^2$ for notational simplicity to denote the number of elements in the matrix.

We can produce an N-by-1 vector of these flows that reflects an origin-centric ordering of the matrix as shown in Table 5.1. The dyad label denotes the overall index from 1, ..., N for the ordering. The vector is formed using $y = vec(Y)$, where $vec(\cdot)$ is an operator that converts a matrix to a vector by stacking the columns as shown in Table 5.1. The first n elements in the stacked vector y reflect flows from origin zone $i = 1$ to all n destinations and the last n elements flows from origin zone $i = n$ to destinations 1, ..., n.

A spatial interaction model in matrix notation relies on two sets of explanatory variable matrices. One is an N-by-Q matrix of Q origin-specific variables A_q ($q = 1, ..., Q$), that we label X_o. This matrix reflects an n-by-Q matrix X of Q explanatory variables that is repeated n times using $X_o = X \otimes \iota_n$, where ι_n is an n-by-1 vector of ones. The matrix Kronecker product $\otimes$ works to multiply the right-hand argument ι_n times each element in the matrix X, which strategically repeats the explanatory variables so they are associated with observations treated as origins. Specifically, the matrix product repeats the origin characteristics of the first location to form the first n rows, the origin characteristics of the second location n times for the next n rows and so on (see Table 5.1), resulting in the N-by-Q matrix X_o (see LeSage and Fischer 2010, p. 414).

The second matrix is an N-by-R matrix $X_d = \iota_n \otimes \tilde{X}$ that represents the R destination-specific characteristics of the n destination locations. The Kronecker product works to repeat the n-by-R matrix $\tilde{X}$ of the R variables B_r ($r = 1, ..., R$) n times to produce an N-by-R matrix representing destination characteristics (see Table 5.1) that we label X_d.

In addition to the two explanatory variable matrices X_o and X_d, an N-by-1 vector $\tilde{D}$ of distances (spatial separations) between each origin–destination dyad is included in the model. This vector is formed, using $\tilde{D} = vec(D)$, where D denotes an n-by-n distance matrix and $vec(\cdot)$ is the operator that converts this distance matrix to an N-by-1 vector by stacking the columns of the matrix, as shown in Table 5.1 (see LeSage and Fischer 2010, p. 415).

Assuming a univariate power function specification of the separation function (see Eq. (4.11)), the log-additive (power deterrence function) spatial

Table 5.1 An origin-centric scheme for origin–destination flow arrangements

Dyad label	ID origin	ID destination	Flows	Origin variables	Destination variables	Distance variable
1	1	1	$y(1,1)$	$A_1(1)...A_Q(1)$	$B_1(1)...B_R(1)$	$D(1,1)$
$\vdots$	$\vdots$	$\vdots$	$\vdots$	$\vdots$ $\vdots$	$\vdots$ $\vdots$	$\vdots$
n	1	n	$y(1,n)$	$A_1(1)...A_Q(1)$	$B_1(n)...B_R(n)$	$D(1,n)$
$n+1$	2	1	$y(2,1)$	$A_1(2)...A_Q(2)$	$B_1(1)...B_R(1)$	$D(2,1)$
$\vdots$	$\vdots$	$\vdots$	$\vdots$	$\vdots$ $\vdots$	$\vdots$ $\vdots$	$\vdots$
$2n$	2	n	$y(2,n)$	$A_1(2)...A_Q(2)$	$B_1(n)...B_R(n)$	$D(2,n)$
$\vdots$	$\vdots$	$\vdots$	$\vdots$	$\vdots$ $\vdots$	$\vdots$ $\vdots$	$\vdots$
$N-n+1$	n	1	$y(n,1)$	$A_1(n)...A_Q(n)$	$B_1(1)...B_R(1)$	$D(n,1)$
$\vdots$	$\vdots$	$\vdots$	$\vdots$	$\vdots$ $\vdots$	$\vdots$ $\vdots$	$\vdots$
N	n	n	$y(n,n)$	$A_1(n)...A_Q(n)$	$B_1(n)...B_R(n)$	$D(n,n)$

interaction model may be written in matrix notation (with all the variables in log-form) as

$$y = \alpha\, \iota_N + X_o\, \beta + X_d\, \gamma - \theta\, \tilde{D} + \varepsilon' \qquad (5.1)$$

with

y N-by-1 vector of origin–destination flows,

X_o N-by-Q matrix of Q origin-specific variables that characterize the ability of the origin locations to produce flows,

β the associated Q-by-1 parameter vector that reflects the origin effects,

X_d N-by-R matrix of R destination-specific variables that represent the attractiveness of the destination locations,

γ the associated R-by-1 parameter vector that reflects the destination effects,

$\tilde{D}$ N-by-1 vector of distances (separations) between origin and destination locations,

θ the associated scalar parameter that reflects the distance (separation) effect,

ι_N N-by-1 vector of ones,

α constant term parameter on ι_N,

ε' N-by-1 vector of normally distributed disturbances with zero mean and constant variances.

This spatial interaction model is based on the independence assumption of origin–destination flows, and called the *independence (log-normal) model of spatial interaction*, since all variables are in log-form. Independence implies that (i) the *individual* flows from origin i to destination j are independent from each other, and that (ii) *aggregate* interaction flows between any pair of locations, say (i, j), are independent from flows between any other pair of locations, say (r, s), with $r \neq i$ and $s \neq j$.

As already pointed out in Sect. 4.3, estimating the parameters $\alpha, \beta,\ \gamma$ and θ by means of OLS would only be justified statistically if we believed that the flows $Y(i,j)$ were independent and log-normally distributed about their mean value with a constant variance. Such an assumption, however, is patently not valid since OD flows are discrete counts whose variance is very likely to be proportional to their mean value. Least squares assumptions ignore the true integer nature of the OD flows and approximate a discrete valued process by an almost certainly misrepresentative continuous distribution (see Bailey and Gatrell 1995, p. 353). Thus, OLS regression estimates and their standard errors can be seriously distorted.

5.2 Econometric Extensions to the Independence Spatial Interaction Model

The simplicity of the log-normal spatial interaction model with independent observations may mean that it cannot account for the spatial richness of origin–destination flows (LeSage and Pace 2008). To enhance this model, we replace the conventional assumption of independence between origin–destination flows with formal approaches that allow for spatial dependence in flow magnitudes.

Specifically, we augment the model with three types of spatial dependence: *origin-based*, *destination-based*, and *origin-to-destination based* dependence. The first type of dependence reflects the notion that forces leading to flows from any origin location to a particular destination may generate similar flows from near-by origin locations. The second type reflects the intuition that forces leading to flows from an origin to a destination location may create similar flows to near-by destination locations. The third type is a combination of origin-based and destination-based dependence (LeSage and Pace 2009, p. 215).

There are two ways to incorporate these types of spatial dependence into the log-normal spatial interaction model given by Eq. (5.1).

The First Approach One way to deal with spatial dependence in origin–destination flows is to specify a (first order) spatial autoregressive process that governs the spatial interaction variable y. This approach leads to a spatial autoregressive extension of the model (LeSage and Pace 2008) in Eq. (5.1).

$$y = \rho_o\, W_o\, y + \rho_d\, W_d\, y + \rho_w\, W_w\, y + \alpha\, \iota_N + X_o\, \beta + X_d\, \gamma - \theta\, \tilde{D} + \varepsilon' \qquad (5.2)$$

$$\varepsilon' \sim N(0,\ \sigma^2 I_N) \qquad (5.3)$$

that takes origin-based, destination-based and origin-to-destination based spatial dependence into account.

The N-by-N spatial weights matrix $W_o = W \otimes I_n$ is used to form a spatial lag vector $W_o\, y$ that captures *origin-based spatial dependence* among flows arising from observations y_{ij} and y_{rj} where origins i and r $(r \neq i)$ represent neighbouring origin locations. The n-by-n spatial weights matrix W is a non-negative matrix with diagonal elements set to zero to prevent a location from being defined as a neighbour to itself. Neighbours could be defined using contiguity or other measures of spatial proximity such as cardinal distance (for example, kilometers) and ordinal distance (for example, the six closest neighbours). See Sect. 2.2 for more details. The spatial weights matrix W is standardised to have row sums of unity, and this is required to produce linear combinations of flows from neighbouring locations in the model given by Eq. (5.2).

Given an origin-centric organisation of the sample data as shown in Table 5.1, the spatial weights matrix $W_o = W \otimes I_n$ will form an N-by-1 vector $W_o y$ containing a linear combination of flows from locations neighbouring each observation treated as an origin. In the case where neighbours are weighted equally, we would have an unweighted average of the neighbouring origin–destination flows.

Similarly, a spatial lag of the dependent variable constructed using the weights matrix $W_d = I_n \otimes W$ to generate an N-by-1 vector $W_d y$, captures *destination-based spatial dependence* using a linear combination of flows associated with observations y_{ij} and y_{is} where j and s $(s \neq j)$ represent neighbouring destination locations. Finally, an N-by-N spatial weights matrix, $W_w = W \otimes W$, can be used to form a spatial lag vector $W_w y$ that captures *origin-to-destination based spatial dependence* using a linear combination of neighbours to both the origin and destination locations (LeSage and Pace 2008).

It is worthwhile to note that model specification (5.2)–(5.3) subsumes different more specific spatial econometric interaction models of interest. These model specifications result from various restrictions on the parameters (see LeSage and Pace 2008): (i) the restriction $\rho_o = \rho_d = \rho_w = 0$ generates the independence log-normal model of spatial interaction given by Eq. (5.1); (ii) the restriction $\rho_d = \rho_w = 0$ results in a spatial interaction model based on the spatial weights matrix W_o, reflecting origin autoregressive spatial dependence; (iii) the restriction $\rho_o = \rho_w = 0$ results in a spatial interaction model based on the weights matrix W_d, reflecting destination autoregressive spatial dependence, and (iv) the restriction $\rho_o = \rho_d = 0$ generates a single weights matrix, W_w, spatial interaction model that accounts for dependence on interaction between origin and destination neighbours (LeSage and Fischer 2010).

The log-likelihood function for the spatial interaction model (5.2) concentrated with respect to the parameters α, β, γ, θ and σ^2 takes the form

$$\ln L_{con}(\rho_o, \rho_d, \rho_w) = \kappa + \ln |I_N - \rho_o W_o - \rho_d W_d - \rho_w W_w|$$
$$- \frac{N}{2} \ln T(\rho_o, \rho_d, \rho_w) \tag{5.4}$$

where $T(\rho_o, \rho_d, \rho_w)$ represents the sum of squared errors expressed as a function of the scalar dependence parameters alone after concentrating out the parameters $\alpha, \beta, \gamma, \theta$ and σ^2, and κ denotes a constant which does not depend on ρ_o, ρ_d, ρ_w (LeSage and Pace 2008).

Model specifications based on a single spatial weights matrix (W_o, W_d or W_w) can be estimated using standard maximum likelihood methods with a numerical Hessian approach to compute estimates of dispersion and t-statistics. Potential computational problems that might plague ML estimation for models with very large N observations can be tackled using specialised approaches suggested, for example, by Smirnov and Anselin (2001); Pace and LeSage (2004), (2009) and LeSage and Pace (2008), to calculate the N-by-N log-determinant $\ln |I_N - \rho_h W_h|$ for $h = o$, d or w.

Estimating the more general models that involve more than a single spatial dependence parameter requires customised algorithms of the type set forth in LeSage and Pace (2008). An applied illustration using state-level population migration flows can be found in this publication as well

The Second Approach Another way to dealing with spatial dependence in origin–destination flows is to specify a spatial process for the disturbance terms, structured to follow a (first order) spatial autoregressive process (see Fischer and Griffith 2008). This specification could be estimated using standard maximum likelihood methods. In this framework, the spatial dependence resides in the disturbance process, as in the case of serial correlation in time series regression models.

Specifically, the most general variant of this type of model specification takes the form (LeSage and Fischer 2010)

$$y = \alpha \, \iota_N + X_o \beta + X_d \gamma - \theta \tilde{D} + u \tag{5.5}$$

$$u = \rho_o \, W_o \, u + \rho_d \, W_d \, u + \rho_w \, W_w \, u + \varepsilon' \tag{5.6}$$

$$\varepsilon' \sim N(0, \ \sigma^2 I_N) \tag{5.7}$$

where the definitions for the spatial lags involving the disturbance terms in Eq. (5.6), $W_o \, u$, $W_d \, u$ and $W_w \, u$, are analogous to those for the spatial lags of the dependent variable in Eq. (5.2).

Simpler models can be constructed by imposing restrictions on the general specification given by Eq. (5.6). For example, we could specify the disturbances using

$$u = \rho \, \tilde{W} u + \varepsilon' \tag{5.8}$$

$$\varepsilon' \sim N(0, \ \sigma^2 I_N) \tag{5.9}$$

which merges origin-based and destination-based spatial dependence to produce a single (row-normalised) spatial weights matrix $\tilde{W}$ consisting of the sum of W_o and W_d which is row-normalised to produce a single vector $\tilde{W} u$ reflecting a spatial lag of the disturbances. This specification also restricts the origin-to-destination-based spatial dependence in the disturbances to be zero, since ρ_w is implicitly set to zero.

The virtue of a simpler model such as that given by Eq. (5.5) along with Eqs. (5.8) and (5.9) is that conventional software for estimating spatial error models could be used to produce an estimate for the parameter ρ along with the remaining model parameters α, β, γ and θ.

Estimating the more general models that involve more than one single spatial dependence parameter requires customised algorithms of the type set forth in LeSage and Pace (2008). These are needed to maximise a log-likelihood that is concentrated with respect to the parameters α, β, γ, θ and σ^2 resulting in an optimisation problem involving more than one of the dependence parameters ρ_o, ρ_d and ρ_w (LeSage and Fischer 2010).

One point to note regarding modelling spatial dependence in the model disturbances is that the coefficient estimates α, β, γ, and θ will be asymptotically equal to those from least squares estimation. But there may be an efficiency gain that arises from modelling dependence in the disturbances.

The Zero Flows Problem   Several problems might arise in applied practice when estimating the spatial econometric interaction models (for a discussion see LeSage and Fischer 2010). Particular attention has to be given to the so-called *zero flows problem* that involves the presence of a large numbers of zero flows. This problem arises when analysing sample data collected using a finer spatial scale.

Zero flows present no serious problem in the case of the Poisson model specifications discussed in the previous chapter, but must be explicitly handled in the case of the log-normal spatial interaction model and its extensions. A large number of zero flows invalidates use of least squares regression as a method for estimating the independence (log-normal) model and maximum likelihood methods for spatial extensions to the independence model. This is because zero values for a large proportion of the dependent variable observations invalidate the normality assumption required for inference in the regression model and validity of the maximum likelihood approach (LeSage and Fischer 2010, p. 427).

Despite this, a number of applications of the independence log-normal model can be found where the dependent variable is modified by adding a small positive number such as 0.5 or one to the observations to accommodate the log-transformation. But this ignores the discrete nature of the flow distribution and, moreover, can have a considerable impact on the coefficients of the model and on its explanatory power (Flowerdew and Aitkin 1982).

5.3   Spatial Filtering Versions of Spatial Interaction Models

Another approach to handling spatial dependence in origin–destination data relies on using Griffith's (2003) spatial filtering methodology, originally developed for area data. This methodology requires the introduction of appropriate synthetic covariates that serve as surrogates for spatially autocorrelated missing origin and destination variables. These synthetic variables are constructed as linear combinations of eigenvectors extracted from the transformed spatial weights matrix W

$$(I_n - \iota_n \, \iota'_n \, n^{-1}) \; W \; (I_n - \iota_n \, \iota'_n \, n^{-1}), \tag{5.10}$$

where $(I_n - \iota_n \, \iota'_n \, n^{-1})$ is a projection matrix, W an n-by-n binary spatial contiguity matrix, I_n the n-by-n identity matrix, and ι_n the n-by-1 vector of ones.

Tiefelsdorf and Boots (1995) show that all of the eigenvalues of expression (5.10) relate to distinct Moran's I values. Matrix (5.10) is of rank $n-1$, and thus has $n-1$ eigenvalues and these are associated with $n-1$ mutually orthogonal and uncorrelated axes with conditionally maximum global spatial autocorrelation measured in terms of Moran's I. The nth eigenvector, say E_n, (whose eigenvalue is zero) is proportional to a vector of ones. The first eigenvector, E_1, measures the maximum global spatial autocorrelation, the second, E_2, measures the maximum residual spatial autocorrelation after extracting the first, and so on. For full details see Tiefelsdorf and Boots (1995).

These n eigenvectors that describe the full range of all possible mutually orthogonal and uncorrelated map patterns may be interpreted as synthetic map variables that represent specific types (that is, positive or negative) and degrees (for example, negligible, weak, moderate, strong) of spatial autocorrelation. So if spatial autocorrelation is present in an analysis, those eigenvectors portraying it can be used to represent it (Griffith 2010).

Operationally, only a subset of the n eigenvectors having I/I_{max} larger than 0.25, is usually needed where I_{max} denotes the maximum possible value of Moran's I statistic. Once eigenvectors have been computed, a stepwise selection procedure can be used to identify statistically significant ones. For further details on eigenvector selection and implementation strategies, see Griffith (2002, 2004), and for a discussion of the relationship between the eigenvectors and the binary contiguity matrix, see Tiefelsdorf and Griffith (2007).

The eigenvector spatial filtering approach, based upon a stepwise selection criterion, adds a minimally sufficient set of statistically significant eigenvectors as proxies for missing origin and destination variables, and in doing so accounts for spatial autocorrelation among the observations by inducing mutual dyad error independence.

The intuition behind the spatial filtering methodology is that we can replace the spatial autoregressive structure that governs the origin-specific and destination-specific effects using an approximation to the spatial autoregressive process by means of an eigenfunction decomposition of the n-by-n binary spatial weights matrix W. That is

$$W = E \Lambda E' \tag{5.11}$$

where E is the n-by-n matrix whose kth column is the basis eigenvector E_p of W and Λ is the diagonal matrix whose diagonal elements are the corresponding eigenvalues w_p. The approximation is accomplished by removing those eigenvectors from the full set according to the above mentioned threshold.

Assuming functional specifications (4.7), (4.8) and (4.14), leads to a spatial filter specification of the spatial interaction model (4.3) given by

$$\mu(i,j) = C \ \exp\left[\sum_{q=1}^{Q'} E_{iq}\psi_q\right](A_i)^{\beta}\ \exp\left[\sum_{r=1}^{R'} E_{jr}\varphi_r\right](B_j)^{\gamma}$$

$$\exp\left[-\sum_{k=1}^{K} \theta_k D^{(k)}(i,j)\right]\quad i,j = 1,\ \dots, n \tag{5.12}$$

where $\mu(i,j)$, $D^{(k)}(i,j)$, A_i, B_j, C, β, γ and θ_k are defined as above (see Chap. 4). Q' and R' denote the number of eigenvectors E_{iq} and E_{jr} selected to furnish a good description of flows out of the origins and flows into the destinations, respectively, and ψ_q and φ_r are the respective coefficients for the linear combinations of eigenvectors that constitute the *origin* and *destination spatial filters*, namely

$$\sum_{q=1}^{Q'} E_{iq}\psi_q \quad i = 1\ \dots, n \tag{5.13}$$

$$\sum_{r=1}^{R'} E_{jr}\varphi_r \quad j = 1\ \dots, n. \tag{5.14}$$

For these spatial filters, which are linear combinations of the eigenvectors and represent the spatial autocorrelation components of the missing origin and destination variables, ψ_q $(q = 1,\ \dots, Q')$ and φ_r $(r = 1,\ \dots, R')$ are regression coefficients that indicate the relative importance of each distinct map pattern in accounting for spatial autocorrelation in the flows structure.

Spatial filter spatial model specification (5.12) can be expressed equivalently in log-additive form as

$$\mu(i,j) = \ln C \ + \sum_{q=1}^{Q'} E_{iq}\psi_q + \beta\ \ln A_i\ + \sum_{r=1}^{R'} E_{jr}\varphi_r + \gamma\ \ln B_j$$

$$-\sum_{k=1}^{K} \theta_k D^{(k)}(i,j) \tag{5.15}$$

in order to link it to the independence (log-normal) model of spatial interaction (see Eq. (5.1)).

OLS can be employed to estimate the parameters of the log-normal additive model of interaction. All conventional diagnostic statistics developed for linear regression analysis can be computed and interpreted without having to develop spatially adjusted counterparts. The major numerical difficulty of the spatial filter model version is that eigenfunctions have to be calculated, a formidable computational task for larger spatial interaction systems (i.e. large n).

It is worth noting that inserting Eq. (5.11) into Eqs. (4.20) and (4.40) yield spatial filter counterparts to the Poisson and the negative binomial spatial interaction model versions described in Sects. 4.4 and 4.6. Parameter estimation can be achieved with maximum likelihood (see Sect. 4.5). The methodology can be implemented in SAS or S-PLUS. An applied illustration using journey-to-work flows among German NUTS-3 regions can be found in Griffith (2009), and another using patent citations among European NUTS-2 regions is given in Fischer and Griffith (2008).

References

Anselin L (2009) Spatial regression. In: Fotheringham AS, Rogerson PA (eds) The SAGE handbook of spatial analysis. Sage, Los Angeles and London, pp 255–276

Anselin L (2006) Spatial econometrics. In: Mills T, Patterson K (eds) Palgrave handbook of econometrics: Econometric theory, vol 1. Palgrave Macmillan, Basingstoke, pp 961–969

Anselin L (2003a) Under the hood. Issues in the specification and interpretation of spatial regression models. Agric Econ 27(3):247–267

Anselin L (2003b) Spatial econometrics. In: Baltagi BH (ed) A companion to theoretical econometrics. Blackwell, Oxford, pp 310–330

Anselin L (1996) The Moran scatterplot as an ESDA tool to assess local instability in spatial association. In: Fischer MM, Scholten HJ, Unwin D (eds) Spatial analytical perspectives on GIS. Taylor & Francis, London, pp 111–125

Anselin L (1995) Local indicators of spatial association–LISA. Geogr Anal 27(2):93–115

Anselin L (1993) Discrete space autoregressive models. In: Goodchild MF, Parks BO, Steyaert ET (eds) Environmental modeling with GIS. Oxford University Press, New York and Oxford, pp 454–469

Anselin L (1988a) Lagrange multiplier test diagnostics for spatial dependence and spatial heterogeneity. Geogr Anal 20(1):1–17

Anselin L (1988b) Spatial econometrics: methods and models. Kluwer, Dordrecht

Anselin L, LeGallo J (2006) Interpolation of air quality measures in hedonic house price models: spatial aspects. Spatial Econ Anal 1(1):31–52

Anselin L, Rey SJ (1991) Properties of tests for spatial dependence in linear regression models. Geogr Anal 23(2):112–131

Anselin L, Syabri I, Kho Y (2010) GeoDa: an introduction to spatial data analysis. In: Fischer MM, Getis A (eds) Handbook of applied spatial analysis. Software tools, methods and applications. Springer, Berlin, pp 73–89

Anselin L, Bera A, Florax RJ, Yoon M (1996) Simple diagnostic tests for spatial dependence. Reg Sci Urban Econ 26(1):77–104

Bailey TC, Gatrell AC (1995) Interactive spatial data analysis. Addison-Wesley Longman, Essex

Barry R, Pace RK (1999) A Monte Carlo estimator of the log determinant of large sparse matrices. Linear Algebra Appl 289:41–54

Baumann J, Fischer MM, Schubert U (1983) A multiregional labour supply model for Austria: the effects of different regionalizations in multiregional labour market modelling. Papers. Reg Sci Assoc 52(1):53–83

Bivand RS (2010) Exploratory spatial data analysis. In: Fischer MM, Getis A (eds) Handbook of applied spatial analysis. Software tools, methods and applications. Springer, Berlin, pp 219–254

Bivand RS, Gebhardt A (2000) Implementing functions for spatial statistical analysis using the R language. J Geogr Syst 2(3):307–317

Bivand RS, Pebesma EJ, Gómez-Rubio V (2008) Applied spatial data analysis with R. Springer, Berlin

Burridge P (1980) On the Cliff-Ord test for spatial autocorrelation. J Royal Stat Soc B 42(1): 107–108

Cameron AC, Trivedi PK (2005) Microeconometrics. Methods and applications. Cambridge University Press, Cambridge

Cameron AC, Trivedi PK (1998) Regression analysis of count data. Cambridge University Press, Cambridge

Cliff AD, Ord JK (1981) Spatial processes: models and applications. Pion, London

Cliff AD, Ord JK (1973) Spatial autocorrelation. Pion, London

Cliff A, Ord JK (1972) Testing for spatial autocorrelation among regression residuals. Geogr Anal 4(3):267–284

Cressie NAC (1993) Statistics for spatial data (revised edition). Wiley, New York, Chichester, Toronto and Brisbane

Demšar U (2009) Geovisualisation and geovisual analytics. In: Fotheringham AS, Rogerson PA (eds) The SAGE handbook of spatial analysis. Sage, Los Angeles and London, pp 41–62

Fingleton B, López-Bazo E (2006) Empirical growth models with spatial effects. Pap Reg Sci 85(2):177–198

Fischer MM (2010) Spatial interaction models. In: Wharf B (ed) Encyclopedia of geography. Sage, London, pp 2645–2647

Fischer MM (2002) Learning in neural spatial interaction models: a statistical perspective. J Geogr Syst 4(3):287–299

Fischer MM (2001) Spatial analysis in geography. In: Smelser NJ, Baltes PB (eds) International Encyclopedia of the Social and Behavioral Sciences, vol 22. Elsevier, Oxford, pp 14752–14758

Fischer MM (2000) Spatial interaction models and the role of geographical information systems. In: Fotheringham AS, Wegener M (eds) Spatial models and GIS. New potential and new models. Taylor & Francis, London, pp 33–43

Fischer MM, Getis A (eds) (2010) Handbook of applied spatial analysis. Software tools, methods and applications. Springer, Berlin

Fischer MM, Griffith DA (2008) Modeling spatial autocorrelation in spatial interaction data: an application to patent citation data in the European Union. J Reg Sci 48(5):969–989

Fischer MM, Reggiani A (2004) Spatial interaction models: from gravity to the neural network approach. In: Cappello R, Nijkamp P (eds) Urban dynamics and growth. Elsevier, Amsterdam, pp 319–346

Fischer MM, Reismann M (2002) A methodology for neural spatial interaction modeling. Geogr Anal 34(3):207–228

Fischer MM, Reismann M, Scherngell T (2006) The geography of knowledge spillovers in Europe–evidence from a model of interregional patent citations in high-tech industries. Geogr Anal 38(3):288–309

Fischer MM, Scherngell T, Jansenberger E (2009a) Geographic localization of knowledge spillovers: evidence from high-tech patent citations in Europe. Ann Reg Sci 43(4):839–858

Fischer MM, Bartkowska M, Riedl A, Sardadvar S, Kunnert A (2009b) The impact of human capital on regional labour productivity in Europe. Lett Spatial Resour Sci 2(2–3):97–108

Florax RJGM, Folmer H (1992) Specification and estimation of spatial linear regression models: Monte Carlo evaluation of pre-test estimators. Reg Sci Urban Econ 22(3):405–432

Florax RJGM, Folmer H, Rey SJ (2003) Specification searches in spatial econometrics: the relevance of Hendry's methodology. Reg Sci Urban Econ 33(5):557–579

Flowerdew R, Aitkin M (1982) A method of fitting the gravity model based on the Poisson distribution. J Reg Sci 22(2):191–202

Fotheringham AS, O'Kelly ME (1989) Spatial interaction models: formulations and applications. Kluwer, Dordrecht

Fortin M-J, Dale MRT (2009) Spatial autocorrelation. In: Fotheringham AS, Rogerson PA (eds) The SAGE handbook of spatial analysis. Sage, Los Angeles and London, pp 89–103

Getis A (2010) Spatial autocorrelation. In: Fischer MM, Getis A (eds) Handbook of applied spatial analysis. Software tools, methods and applications. Springer, Berlin, pp 255–278

Getis A (1995) The tyranny of data. Tenth University Research Lecture, San Diego State University. San Diego State University Press

Getis A, Griffith DA (2002) Comparative spatial filtering in regression analysis. Geogr Anal 34(2):130–140

Getis A, Ord JK (1996) Local spatial statistics: an overview. In: Longley P, Batty M (eds) Spatial analysis: modelling in a GIS environment. John Wiley & Sons, New York, pp 261–277

Getis A, Ord JK (1992) The analysis of spatial association by distance statistics. Geogr Anal 24(3):189–206

Griffith DA (2010) Spatial filtering. In: Fischer MM, Getis A (eds) Handbook of applied spatial analysis. Software tools, methods and applications. Springer, Berlin, pp 301–318

Griffith DA (2009) Modeling spatial autocorrelation in spatial interaction data: empirical evidence for 2002 Germany journey-to-work flows. J Geogr Syst 11(2):117–140

Griffith DA (2007) Spatial structure and spatial interaction: 25 years later. Rev Reg Stud 37(1):28–38

Griffith DA (2004) Distributional properties of georeferenced random variables based on the eigenfunction spatial filter. J Geogr Syst 6(3):263–288

Griffith DA (2003) Spatial autocorrelation and spatial filtering. Springer, Berlin

Griffith DA (2002) A spatial filtering specification for the auto-Poisson model. Stat Probab Lett 58(2):245–251

Griffith DA (2000) A linear regression solution to the spatial autocorrelation problem. J Geogr Syst 2(2):141–156

Griffith DA (1988) Advanced spatial statistics. Kluwer, Dordrecht

Griffith DA, Sone A (1995) Trade-offs associated with normalizing constant computational simplifications for estimating spatial statistical models. J Stat Comput Simul 51(2–4):165–183

Haining RP (2010) The nature of georeferenced data. In: Fischer MM, Getis A (eds) Handbook of applied spatial analysis. Software tools, methods and applications. Springer, Berlin, pp 197–217

Haining RP (2003) Spatial data analysis: theory and practice. Cambridge University Press, Cambridge

Haining RP (1990) Spatial data analysis in the social and environmental sciences. Cambridge University Press, Cambridge

Haining R, Law J, Griffith D (2009) Modelling small area counts in the presence of overdispersion and spatial autocorrelation. Comput Stat Data Anal 53(8):2923–2937

Kelejian H, Prucha IR (2010) Spatial models with spatially lagged dependent variables and incomplete data. J Geogr Syst 12(3):241–257

Kelejian H, Prucha IR (1999) A generalized moments estimator for the autoregressive parameter in a spatial model. Int Econ Rev 40(2):509–533

Kelejian H, Prucha IR (1998) A generalized spatial two stage least squares procedure for estimating a spatial autoregressive model with autoregressive disturbances. J Real Estate Finance Econ 17(1):99–121

Kelejian H, Tavlas GS, Hondronyiannis G (2006) A spatial modeling approach to contagion among emerging economies. Open Econ Rev 17(4/5):423–442

Kim CW, Phipps TT, Anselin L (2003) Measuring the benefits of air quality improvement: a spatial hedonic approach. J Environ Econ Manag 45(1):24–39

Ledent J (1985) The doubly constrained model of spatial interaction: a more general formulation. Environ Plan A 17(2):253–262

Lee M, Pace RK (2005) Spatial distribution of retail sales. J Real Estate Finance Econ 31(1):53–69

LeSage JP (1997) Bayesian estimation of spatial autoregressive models. Int Reg Sci Rev 20(1/2): 113–129

LeSage JP, Fischer MM (2010) Spatial econometric methods for modeling origin-destination flows. In: Fischer MM, Getis A (eds) Handbook of applied spatial analysis. Software tools, methods and applications. Springer, Berlin, pp 413–437

LeSage JP, Fischer MM (2008) Spatial growth regressions: model specification, estimation and interpretation. Spatial Econ Anal 3(3):275–304

LeSage JP, Pace RK (2010) Spatial econometric models. In: Fischer MM, Getis A (eds) Handbook of applied spatial analysis. Software tools, methods and applications. Springer, Berlin, pp 355–376

LeSage JP, Pace RK (2009) Introduction to spatial econometrics. CRC Press (Taylor & Francis Group), Boca Raton

LeSage JP, Pace RK (2008) Spatial econometric modeling of origin-destination flows. J Reg Sci 48(5):941–967

LeSage JP, Pace RK (2004) Introduction. In: LeSage JP, Pace RK (eds) Spatial and spatiotemporal econometrics. Elsevier, Amsterdam, pp 1–32

LeSage JP, Fischer MM, Scherngell T (2007) Knowledge spillovers across Europe: evidence from a Poisson spatial interaction model with spatial effects. Pap Reg Sci 86(3):393–421

Liu X, LeSage JP (2010) Arc_Mat: a Matlab-based spatial data analysis toolbox. J Geogr Syst 12(1):69–87

Longley PA, Goodchild MF, Maguire DJ, Rhind DW (2001) Geographic information systems and science. Wiley, Chichester

Martin RJ (1993) Approximations to the determinant term in Gaussian maximum likelihood estimation of some spatial models. Communications in Statistics: theory and Methods 22(1): 189–205

Mur J, Angulo A (2006) The spatial Durbin model and the common factor tests. Spatial Econ Anal 1(2):207–226

Openshaw S (1981) The modifiable areal unit problem. University of East Anglia, Norwich

Openshaw S, Taylor P (1979) A million or so correlation coefficients: the experiments on the modifiable areal unit problem. In: Wrigley N (ed) Statistical applications in the spatial sciences. Pion, London, pp 127–144

Ord JK (1975) Estimation methods for models of spatial interaction. J Am Stat Assoc 70(1): 120–126

Ord JK, Getis A (1995) Local spatial autocorrelation statistics: distributional issues and an application. Geogr Anal 27(4):286–306

Pace RK, Barry RP (1997) Quick computation of spatial autoregressive estimates. Geogr Anal 29(3):232–246

Pace RK, LeSage JP (2009) A sampling approach to estimate the log determinant used in spatial likelihood problems. J Geogr Syst 11(3):209–225

Pace RK, LeSage JP (2006) Interpreting spatial econometric models. Paper presented at the RSAI North-American meetings, Toronto, Nov 2006

Pace RK, LeSage JP (2004) Techniques for improved approximation of the determinant term in the spatial likelihood function. Comput Stat Data Anal 45(2):179–196

Ripley B (1988) Statistical inference for spatial processes. Cambridge University Press, Cambridge

Roy JR (2004) Spatial interaction modelling. A regional science context. Springer, Berlin

Roy JR, Thill J-C (2004) Spatial interaction modelling. Pap Reg Sci 83(1):339–361

Sen A, Smith TE (1995) Gravity models of spatial interaction behavior. Springer, Berlin

Sidák Z (1967) Rectangular confidence regions for the means of multivariate normal distributions. J Am Stat Assoc 62(318):626–633

Smirnov O, Anselin L (2001) Fast maximum likelihood estimation of very large spatial autoregressive models: a characteristic polynomial approach. Comput Stat Data Anal 35(3):301–319

Tiefelsdorf M (2003) Misspecification in interaction model distance relations: a spatial structure effect. J Geogr Syst 5(1):25–50

Tiefelsdorf M, Boots B (1995) The specification of constrained interaction models using SPSS loglinear procedure. Geogr Syst 2(1):21–38

Tiefelsdorf M, Griffith DA (2007) Semiparameter filtering of spatial autocorrelation: the eigenvector approach. Environ Plan A 39(5):1193–1221

Tobler WR (1970) A computer movie simulating urban growth in the Detroit region. Econ Geogr 46(2):234–240

Tukey JW (1977) Exploratory data analysis. Addison Wesley, Reading

Upton G, Fingleton B (1985) Spatial data analysis by example. Wiley, New York

Wang JF, Haining R, Cao ZD (2010) Sampling surveying to estimate the mean of a heterogeneous surface: reducing the error variance through zoning. Int J Geogr Inform Sci 24(4):523–543

Wilson AG (1967) A statistical theory of spatial distribution models. Transp Res 1(3):253–269

Author Index

M. M. Fischer and J. Wang, *Spatial Data Analysis*,
SpringerBriefs in Regional Science, DOI: 10.1007/978-3-642-21720-3,
© Manfred M. Fischer 2011

Subject Index

A

Approximate sampling
distribution, 24–25
Area data, 1, 4–5, 10–44
Area objects
irregularly shaped, 5–7
regularly shaped, 5–7
Areal units. *See* Area objects
Assigning locations to spatial
objects, 6

B

Bishop (spatial) contiguity, 8

C

Cartogram, 18–19
Centroid, 6
Choropleth map, 16–17
Classes of spatial data, 3–5
Class interval selection
equal interval breaks, 16
natural breaks, 16
quantile breaks, 16
standard deviation classifications, 16
Classical linear regression model, 32
Cold spots, 26
Concentrated log-likelihood function for
the SAR and the SDM models, 39
the spatial econometric interaction
models, 64–67
Connectivity matrix. *See* Spatial weights
matrix
Contiguity-based spatial weights
bishop contiguity, 8

first order contiguity. *See* Spatial
weights matrix, first order
contiguity
higher order contiguity. *See* Spatial
weights matrix, higher order
contiguity
queen contiguity, 8
rook contiguity, 8
second order contiguity. *See* Spatial
weights matrix, second order
contiguity
Conventional regression model. *See* Classical
linear regression model
Cross-regressive model, 32

D

Data, 1–2
Data generating process,
spatial autoregressive model, 41
Destination-based spatial dependence, 64–66
Destination-constrained case of spatial inter-
action, 50
Destination-specific factors (variables) in
spatial interaction models, 49–52
Deterrence function. *See* Spatial interaction
models, spatial separation functions
Distance-based spatial weights, 20

E

Eigenfunction (based) spatial filtering, 67–69
Equidispersion property of the Poisson distri-
bution, 54
ESDA. *See* Exploratory spatial data analysis
Exploratory spatial data analysis, 13, 15, 18

M. M. Fischer and J. Wang, *Spatial Data Analysis*,
SpringerBriefs in Regional Science, DOI: 10.1007/978-3-642-21720-3,
© Manfred M. Fischer 2011

MIX
Papier aus verantwortungsvollen Quellen
Paper from responsible sources
FSC® C105338

If you have any concerns about our products,
you can contact us on
ProductSafety@springernature.com

In case Publisher is established outside the EU,
the EU authorized representative is:
Springer Nature Customer Service Center GmbH
Europaplatz 3, 69115 Heidelberg, Germany

Printed by Libri Plureos GmbH
in Hamburg, Germany